Mapping Urban Environmental Risks Using Open Source GIS

Cities across the world are facing a convergence of environmental risks, from flooding and extreme heat events to air pollution and water scarcity. These threats are increasingly uneven in their impacts and deeply shaped by local geography. This book offers a practical, accessible guide to understand and respond to these threats using open source geographic information systems (GIS) and freely available spatial data. Through real-world case studies, the author discusses each risk separately to meet the needs of decision-making when urban threats happen.

Features:

- Focuses on mapping specific urban threats such as water scarcity, flood risk, fire, and urban heat islands.
- Takes a risk-by-risk approach grounded in the everyday realities of each city and uses geospatial tools to support action.
- Shows how to analyze urban threats using open tools like QGIS, freely accessible satellite imagery, and public datasets.
- Emphasizes concepts and decision-making rather than software mastery, making it suitable for a variety of readers.

By demystifying geospatial analysis and emphasizing hands-on, place-based projects, this book empowers readers, such as students, city planners, journalists, community advocates, and public health officials alike, to investigate environmental risks in their own neighborhoods and use maps to support smarter, more equitable action.

David R. Rain is Associate Professor of Geography and International Affairs at the George Washington University. He is a human-environment geographer with interests in the health and environmental consequences of international urbanization, geospatial science and technology, population and census geography, disaster vulnerability and resilience, and geographic and environmental education. He is the author of *Eaters of the Dry Season* and the *UN Handbook on Geospatial Infrastructure in Support of Census Activities.*

GIS in Action

Series Editor: Jochen Albrecht

Geographic Information Systems (GIS) are a cross-disciplinary, integrative technology widely used by professionals around the world in evidence-based spatial decision-making. GIS is application-driven, concerned with the effective use and management of data, empowering its spatial analysis, visualization, and dissemination. This book series covers the disciplinary intersections that happen when GIS is used in action to solve or manage real issues in society. The range of GIS applications includes archaeology, agriculture, environmental and civil engineering, meteorology and climatology, public health, emergency planning and response, forestry, geology, spatial data infrastructures, green energy, transport logistics, and many more. The books in this series address the relevant practical and technological issues, provide case studies, and discuss the impact that applications have on policy and society. It provides academics and professionals the opportunity to combine their research, consultancy, and relevant project work into a book that is then germane to professional lines of work or teaching modules focusing on specific GIS applications.

GIS and Housing: Principles and Practices
Laxmi Ramasubramanian, Jochen Albrecht, and Deborah Rojas De Leon

Geospatial Information to Support the UN's Sustainable Development Goals
Edited by Etta D. Jackson and Amor Laaribi

Mapping Urban Environmental Risks Using Open Source GIS
David R. Rain

For more information about the series, please visit: https://www.routledge.com/GIS-in-Action/book-series/CRCGISINACTI

Mapping Urban Environmental Risks Using Open Source GIS

David R. Rain

CRC Press
Taylor & Francis Group
Boca Raton London New York

CRC Press is an imprint of the
Taylor & Francis Group, an **informa** business

Designed cover image: © Pexels, Franco Monsalvo

First edition published 2026
by CRC Press
2385 NW Executive Center Drive, Suite 320, Boca Raton FL 33431

and by CRC Press
4 Park Square, Milton Park, Abingdon, Oxon, OX14 4RN

CRC Press is an imprint of Taylor & Francis Group, LLC

ISBN: 9781032636924 (hbk)
ISBN: 9781032636931 (pbk)
ISBN: 9781032636948 (ebk)

DOI: 10.1201/9781032636948

Typeset in Palatino
by codeMantra

Contents

Preface: Death by Dashboard

This book is an argument for the public value of geospatial data—and a warning about how easily that value can be lost.

Over the past two decades, cities around the world have become saturated with dashboards. Interactive maps, indicators, and visual summaries promise clarity, transparency, and evidence-based decision-making. They circulate through donor reports and conference presentations, then linger on agency websites long after priorities have shifted. Many of these systems fail to shape outcomes on the ground. They are commissioned, delivered, briefly admired, and quietly sidelined. The data behind them age quickly, slip out of reach, or disappear behind proprietary systems, while the people most exposed to environmental risk rarely see—let alone control—the information meant to protect them.

This pattern, which I refer to as "death by dashboard," is not a failure of geospatial technology. It is a failure of governance and imagination. Dashboards are too often treated as endpoints rather than instruments, serving as symbols of accountability instead of tools for action. Once a contract is fulfilled or a pilot is completed, responsibility dissipates. The interface persists, but the problem remains.

A familiar scene helps explain how this happens. In a national development project in East Africa, a government ministry issues a competitive tender for a geospatial system intended to support land-use planning and environmental risk management. Bids arrive from a mix of small local organizations, often operating with limited staff and equipment, and large international consulting firms with polished proposals and global portfolios. On paper, the process appears open and competitive. In practice, the outcome is rarely in doubt.

The winning bidder is often already known. It is a firm represented by someone who understands how these systems work, not the software, but the institutions that surround procurement, reporting, and delivery. He has the right accent, the right school ties, and a record of projects that satisfy contractual requirements. There is no need to question his competence. The system is delivered on time and under budget. The dashboard functions as specified. And then the project ends.

The data remain difficult to access. The methods are poorly documented. Local agencies lack the authority or resources to adapt the system once consultants depart. The dashboard survives as an artifact of success, while the underlying environmental risks continue to unfold. Accountability has been satisfied procedurally, but capacity has not been built.

This is not corruption in the narrow sense. It is a structural failure. Procurement systems reward delivery rather than durability. Funding cycles privilege visibility over stewardship. Spatial information is produced without clear pathways into decision-making, and responsibility is obscured rather than clarified.

It would be easy to mistake the current moment for one of geospatial abundance. Satellite imagery is cheaper and more frequent than at any point in history. Sensors are widely deployed. Open data portals advertise thousands of downloadable layers. Interactive maps populate government websites and humanitarian platforms. By almost any practical measure, there is more spatial data available today than ever before.

Yet availability is not the same as accessibility, and access is not the same as agency.

Much of what appears to be open is difficult to use in practice. Data are released without adequate documentation or lack the metadata needed for interpretation. Other datasets are nominally public but functionally enclosed, accessible only through proprietary platforms or licensing arrangements that change without notice. The result is a widening gap between those who can view spatial data and those who can act on it.

This gap is especially pronounced in cities in the Global South. Environmental risk is intensifying at the same moment that public geospatial capacity is being fragmented. Floods, heat, air pollution, wildfire, and infrastructure failure are all spatial threats, yet the ability to analyze them increasingly depends on access to tools and data that are unevenly distributed. This reality stands in tension with repeated global commitments to evidence-based planning and climate adaptation.

One reason geospatial projects so often stall at the point of action is that data do not circulate through neutral or coherent pipelines. Spatial information is produced, governed, and contested across complex institutional landscapes shaped by mandates, funding streams, legal authority, and technical capacity. Authority, access, and accountability are distributed unevenly. Some actors control base data but lack analytical capacity, while others possess advanced tools but a limited mandate. Understanding this landscape is a prerequisite for action, and this book returns to it repeatedly.

If the problem were simply a lack of data or computing power, the solution would be straightforward: build bigger systems, commission more dashboards, and invest in ever more sophisticated platforms. That logic has dominated geospatial practice for years, particularly in development and environmental governance, and it has failed repeatedly.

The alternative proposed in this book is deliberately modest: Be a David, not a Goliath.

This is not a call to reject technology or scale, but to reject the assumption that scale, capital, and complexity are prerequisites for impact. Repeatedly, meaningful gains in urban resilience have emerged from small teams working with open data, transparent methods, and a clear understanding of local conditions. These efforts succeed not because they are comprehensive, but because they are targeted.

Large centralized geospatial systems often struggle under their own weight, requiring specialized expertise and institutional stability that many cities do not have. When political priorities shift or contracts end, these systems decay quickly. Smaller, more agile approaches can adapt. They can be maintained by local participants, making it easier to embed spatial reasoning in everyday decision-making rather than confining it to technical units or external consultants.

Being a David also means rethinking what counts as expertise. Effective spatial analysis does not belong exclusively to professional GIS specialists or well-funded agencies. Concerned citizens, planners, journalists, community organizations, and public officials can all use geospatial data to interrogate claims, expose risk, and advocate for change. What matters is the ability to ask spatial questions clearly and to evaluate data critically.

This orientation is especially important in cities in the Global South, where institutional capacity is often constrained and environmental threats are acute. Waiting for perfect data or ideal governance conditions is rarely an option. Action must proceed with what is available, while remaining attentive to uncertainty and inequality.

This book is not a catalog of geospatial software, nor is it a showcase of the latest platforms or visualizations. It does not assume that better interfaces or more automation will, by themselves, lead to better outcomes. Instead, it is about using spatial thinking to take action.

The chapters that follow move from foundational ideas to applied cases. Early chapters introduce core geospatial concepts, open data sources, and ways of thinking about scale and uncertainty. Subsequent chapters examine specific urban environmental threats, treated not only as physical processes but also as spatially distributed challenges shaped by governance and power.

Throughout, maps are treated as arguments rather than artifacts. They are tools for framing problems, revealing trade-offs, and making claims that can be debated in public. Geospatial analysis matters most when it enters the arenas where choices are made: in planning offices, courtrooms, newsrooms, and community meetings.

This book is written for readers who want to act. Students, planners, journalists, advocates, and public officials all encounter environmental risks in their work and daily lives. The intention here is not to prescribe solutions alone, but to provide a framework for using spatial evidence responsibly and effectively. What follows is a guide to thinking spatially about danger—not simply to document it, but to change how decisions are made.

Acknowledgments

This book would not be possible without the friendly cooperation and support of many people. Gratitude goes to my colleagues, staff, and students in the Department of Geography and Environment at The George Washington University. My colleagues at the Institute for International Economic Policy in the Elliott School, including James Foster, Stephen Smith, Remy Jedwab, Kyle Renner, Alex Rivera, and ESIA Dean Alyssa Ayres, also provided much-appreciated extended support. Students from the Humanitarian Mapping Society student organization and Youth Mappers continue to provide inspiration for human betterment through mapping. For this book, the cartographic artistry and technical help of Maya O'Brien and John Lesko were indispensable. In addition, undergraduate GW students Celeste Robertson, Katherine Petrillo, Joseph Bonanno, and Esther Ki provided essential research assistance along the way. My long-time friend and colleague at the UN Statistics Division, Amor Laaribi, opened the door for this book. Esther Onyekachi Moore, Kenneth Mubea, Pauline Okeyo, Godwin Arku, Joseph Kerski, Primoz Kovacic, and Sulochana Shekhar reviewed the proposal and gave useful feedback. Editor Irma Britton provided me with much-appreciated extra time to get through the process. Special thanks go to my family—front and center my Ollie Street pod, partner Anna, and son Jasper—and my other adult children, Avery, Lydia, and Callen, their respective partners Kayla, Joe, and Sarah, and grandchildren Sylvie, Hugo, Delia, and Mabel. My late parents, Phyllis and Don Rickter, listened to my stories and underwrote my wanderlust. It is their legacy that makes it possible to offer this book as an open-access publication. I appreciate my brother Paul and spouse Ellen for being a landing place for me and a sounding board for so much aging parent content. Lastly, and ultimately, this book is dedicated to mappers, geographers, teachers, journalists, activists, artists, dancers, nerds, techies, engineers, visual thinkers, and the curious of every stripe.

1

Introduction: How to Use This Book

Chapter Introduction

This book is a practical guide to mapping urban environmental threats and using geospatial tools to support action. It is written for a broad audience, including students, planners, researchers, community advocates, journalists, civic technologists, and public officials—particularly those working in or with cities in low- and middle-income regions. Throughout, the emphasis is on concepts rather than code. While examples reference specific tools, including open-source platforms such as QGIS, the methods described here are not software dependent. The goal is to help readers think spatially, evaluate data critically, and design analyses that can inform real decisions. Maps are treated not as technical endpoints, but as arguments used to frame problems, test claims, and support action.

Urban environmental risks are not abstract phenomena. They are experienced unevenly across neighborhoods, shaped by infrastructure, governance, and historical patterns of investment and exclusion. Geospatial analysis offers a way to make these patterns visible and contestable. When used well, spatial tools help structure questions, reveal trade-offs, and clarify where interventions may have the greatest effect. When used poorly, they can obscure responsibility or reduce complex problems to static visuals. This book is written with the conviction that the difference lies not in the sophistication of the software, but in how spatial thinking is applied.

1.1 How the Book is Organized

The book moves from foundational geospatial ideas to applied analyses of specific urban environmental threats. Chapters build on one another conceptually, but they are designed to stand alone for teaching, reference, or applied project design. Readers may move sequentially through the book or engage selectively with chapters relevant to their interests or work.

The opening chapters establish shared ground. Chapter 2 reviews core GIS concepts, with an emphasis on open data, transparency, and reproducible

DOI: 10.1201/9781032636948-1

workflows. Chapter 3 introduces the Digital Elevation Model (DEM) as a foundational dataset for understanding terrain, drainage, and urban form—features that shape many environmental risks, from flooding and landslides to heat and air quality.

Chapters 4–11 examine specific threats, including water access, flooding, fire, heat, wind, landslides, air pollution, and public health. Each of these chapters follows a common analytical structure. They begin with a narrative grounded in a real place, situating the threat in lived experience rather than abstraction. They then examine the physical and social dynamics that produce risk, before framing the problem geospatially and demonstrating how spatial analysis can inform action. The intent is not to provide exhaustive technical instruction, but to show how spatial evidence can be assembled, interpreted, and mobilized in practice.

The final chapter, *Mapping Hope*, brings these threads together by focusing explicitly on resilience and civic engagement. Rather than treating mapping as a diagnostic exercise alone, it emphasizes how geospatial tools can support design, participation, and accountability, moving from problem identification toward implementation.

1.2 Getting Started: A Practical Orientation

Readers come to this book with different backgrounds and goals. Some may be encountering geospatial analysis for the first time, while others already work with maps but seek clearer guidance on applying spatial tools to environmental problems under real-world constraints. Still others may be searching for ways to translate data into action in institutional settings where time, authority, and resources are limited.

Regardless of background, a productive way to approach this book is to think in terms of questions rather than tools. Each chapter begins with a concrete problem grounded in a specific place and encourages readers to ask a small set of recurring questions: What is happening here? Who is affected, and how unevenly? What decisions are being made, and by whom? The role of geospatial analysis is not to automate answers, but to structure these questions spatially and make assumptions explicit.

The chapters do not assume access to advanced software, high-performance computing, or proprietary datasets. Many examples rely on widely available data sources and relatively simple analytical techniques. Where more advanced approaches are introduced, they are framed conceptually, with attention to why they are useful and what their limitations are, rather than as step-by-step technical instructions. This approach reflects the reality that many readers work in contexts where technical capacity is uneven and must be built incrementally.

Different readers may use the chapters in different ways. Those working on real projects may find it useful to skim a chapter first, then return to specific sections as needed. Instructors may assign chapters selectively, using the narrative openings and project design questions to frame discussion and treating the lab or modeling sections as optional extensions. Practitioners may move directly to the geospatial framing and action-oriented sections to see how similar problems have been approached elsewhere. The book is designed to support this flexibility while maintaining a consistent analytical logic across chapters.

1.3 Guiding Principles

Several guiding principles shape the approach taken throughout this book. These are not abstract commitments, but practical constraints that influence how analyses are designed, interpreted, and used.

Transparency is foundational. Spatial analyses should be understandable to non-specialists and replicable by others. Methods matter as much as results, particularly when maps inform public decisions. Wherever possible, the book favors open data, clearly documented workflows, and analytical choices that can be explained without reference to proprietary tools or black-box models. Transparency is treated not as a technical preference, but as a condition for accountability.

Equity is central to how environmental risk is framed. Urban threats are unevenly distributed across space and population, shaped by income, race, infrastructure, and governance. Mapping should make these disparities visible rather than obscure them. This includes attention to who is exposed, who benefits from interventions, and whose data are missing or unreliable. Throughout the book, spatial analysis is used to surface inequality, not to normalize it.

Actionability guides how maps are evaluated. Maps are treated as instruments for decision-making rather than as final products. Each chapter asks how spatial analysis can inform specific choices, such as where to invest, whom to protect first, or how to evaluate alternatives. Analyses that cannot plausibly connect to action are treated as incomplete, regardless of their technical sophistication.

Interdisciplinarity reflects the nature of urban environmental problems themselves. These threats emerge at the intersection of physical processes, social conditions, and governance systems. Effective spatial analysis therefore requires integrating terrain, infrastructure, demographics, institutions, and policy. No single dataset or disciplinary lens is sufficient on its own, and the book emphasizes the value of combining perspectives rather than privileging any one approach.

Durability shapes methodological choices. The most valuable analyses are those that can be updated, reused, and maintained locally over time. This book prioritizes approaches that rely on stable data sources, modest technical requirements, and workflows that can survive beyond a single project or funding cycle. Durability is treated as a form of resilience, particularly in contexts where institutional capacity is constrained.

1.4 How to Use This Book

Readers may move through the chapters sequentially or selectively, depending on their goals. Instructors may adapt chapters as teaching modules or laboratories, while practitioners may draw on specific workflows or case studies relevant to their city or region. Examples are drawn from multiple parts of the world to emphasize comparison rather than prescription. While many cases focus on cities in low- and middle-income regions, the methods apply broadly. Environmental threats do not respect income categories or national borders, and neither do the spatial tools used to analyze them.

The chapters assume a basic familiarity with maps and spatial data, but not advanced technical expertise. Concepts are introduced with an emphasis on intuition, interpretation, and practical relevance rather than software mastery. The intent is to support readers in asking better spatial questions, even when data are incomplete or conditions are imperfect.

Ultimately, this book is an invitation to look at familiar places differently, to question how risk is produced and distributed, and to use spatial evidence not only to understand environmental threats but also to argue for safer and more equitable urban futures.

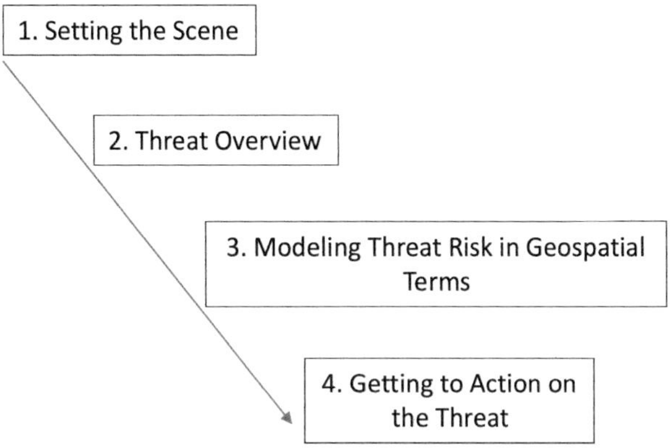

FIGURE 1.1
Analytical workflow schematic diagram.

As depicted in Figure 1.1, each chapter follows a common analytical workflow, beginning with a narrative grounded in a real place, moving through an explanation of the threat and its drivers, applying geospatial analysis, and concluding with a discussion of how spatial evidence can inform action. The workflow emphasizes iteration rather than linear problem-solving, encouraging readers to revisit earlier stages as conditions, data, or priorities change.

2

GIS Foundations in Action

Chapter Introduction

Geographic information systems are often introduced as collections of tools: layers, projections, databases, and software environments that allow users to store and visualize spatial data. While these components matter, they are not sufficient to explain how GIS functions in practice, particularly when the goal is to understand and reduce environmental risk in real places.

This chapter treats GIS foundations as a form of applied reasoning rather than a technical inventory. It traces the intellectual lineage of GIS, explains core spatial data models and analytical operations, and connects those concepts to contemporary practice using open tools and open data. The emphasis throughout is on producing spatial analyses that are transparent, interpretable, and durable, even under the constraints that characterize much urban environmental work.

2.1 From Cartographic Overlay to Digital Analysis

Historically, GIS emerged from the convergence of cartography and database management. Early spatial analysis relied on manual overlays of tracing paper, a technique widely used in World War II and later popularized in the 1960s by landscape architect Ian McHarg (1969) in *Design with Nature.* By stacking transparent layers representing soil, vegetation, hydrology, and land use, analysts could reason about suitability and risk long before computers were involved.

By the 1970s and 1980s, advances in mainframe and minicomputing made it possible to encode spatial features and their attributes digitally, enabling automated overlay, querying, and measurement at scales that manual cartographic methods could not support. Early national mapping and census efforts—most notably the Canada Geographic Information System—demonstrated that geographic data could function as a computational database rather than a static map. These developments coincided with the emergence of topological data models, which formalized spatial relationships

DOI: 10.1201/9781032636948-2

such as adjacency, connectivity, and containment, laying the groundwork for network analysis and infrastructure modeling. Commercial systems such as Arc/Info consolidated these ideas into a coherent analytical framework during the 1980s, while the 1990s saw GIS move from centralized computing environments to desktop platforms through tools like MapInfo and ArcView.

In parallel, satellite remote sensing—driven in part by military and intelligence applications—expanded the scale and frequency of spatial observation, forcing GIS to integrate raster imagery with vector data. By the early 2000s, standardized data formats, coordinate systems, and web-based map services enabled geographic information to circulate widely, setting the stage for platform-based mapping systems such as Google Earth and Google Maps. Contemporary GIS builds on this lineage, combining spatial geometry, attribute data, and analytical functions within software environments that support hypothesis testing, scenario modeling, and public communication. The logic of overlay remains central, but it is now embedded in infrastructures that shape how urban problems are framed, governed, and acted upon.

2.2 Why GIS Aligned So Naturally with the State

Modern GIS also evolved in close alignment with the administrative needs of the state, particularly through census modernization efforts in the United States and elsewhere. Long before "GIS" became a commercial category, national statistical agencies were already confronting problems that were inherently geographic: enumerating populations, defining administrative boundaries, allocating representation, and distributing public resources. The decennial census required stable yet flexible spatial units—tracts, blocks, and enumeration areas—that could persist across survey cycles while adapting to urban growth and political change. Encoding these units digitally transformed geography into a database problem rather than a cartographic one, demanding computational approaches capable of managing topology, address matching, and aggregation across nested spatial scales.

A pivotal moment in this transition was the development of national street-based geographic frameworks within the U.S. Census Bureau. Early work on Dual Independent Map Encoding (DIME) in the 1970s introduced topological street encoding to support address matching and urban enumeration. This approach was later extended and unified at national scale through the Topologically Integrated Geographic Encoding and Referencing (TIGER) system, developed during the 1980s and deployed for the 1990 Census (Marx, 1990). TIGER linked street centerlines, address ranges, and administrative boundaries into a single digital framework, enabling large-scale geocoding and routing well beyond its original census applications. From this perspective, the close relationship between GIS vendors, government agencies, and defense and intelligence organizations is not incidental. The core logic of GIS—overlaying layers, enforcing boundary consistency, linking attributes to place, and summarizing patterns across jurisdictions—mirrors the operational logic of governance itself. As commercial platforms emerged and matured, they inherited both the technical architecture and the institutional priorities of these public-sector origins. GIS thus became embedded not only in software environments, but also in the routine ways that states observe territory, define populations, and intervene in urban systems.

This legacy matters today because the spatial data that are easiest to acquire, most regularly updated, and most finely resolved still tend to reflect state priorities—leaving informal settlements, marginal populations, and non-administrative forms of risk unevenly visible or entirely absent.

2.3 Open Data and the Geography of Access

Open geospatial data are often presented as a technical achievement: a triumph of storage capacity, bandwidth, and interoperability. Over the past two decades, governments, international organizations, and private firms have released unprecedented volumes of spatial data—satellite imagery, administrative boundaries, demographic indicators, and environmental measurements—often framed as evidence of transparency and modernization. Open data portals have become a standard feature of digital government and development practice, and geospatial information is routinely invoked as an enabler of evidence-based planning and sustainable development.

Yet the presence of data does not guarantee its usefulness, and openness does not ensure agency. A central lesson of recent experience is that availability, accessibility, and authority are not the same thing. Data may be technically open but practically unusable: released without adequate documentation, stripped of metadata, published in unstable formats, or hosted on platforms that change licensing terms without warning. In other cases, data are visible but not durable. They are available only for the lifespan of a project or funding cycle, then quietly removed or archived once institutional priorities shift. These conditions limit who can meaningfully use spatial data and for what purposes.

The open data movement itself emerged as a response to earlier regimes of enclosure, where public information was tightly controlled by state agencies or commercial vendors (Chignard, 2013). In the geospatial domain, open standards, open software, and open datasets promised to democratize access to spatial evidence and reduce dependence on proprietary systems.

These efforts have delivered real gains. Today, it is possible to build credible spatial analyses using entirely open-source tools and publicly available data—an outcome that would have been unthinkable a generation ago.

At the same time, new forms of enclosure have emerged. High-resolution satellite imagery, real-time sensor feeds, and advanced analytics are increasingly controlled by private firms operating under opaque licensing regimes. Public agencies often depend on vendor-managed platforms to access their own data, effectively renting analytical capacity rather than owning it. When contracts end, access can disappear. Capacity is not transferred; it is temporarily leased.

Critics within the aid and development community have noted that these arrangements can reproduce inequality rather than reduce it. When local agencies lack the resources to sustain licenses or retain data after projects conclude, the benefits of geospatial investment accrue elsewhere (Gayton, 2020; Mutinda, 2023). This dynamic is especially pronounced in cities in the Global South, where environmental risks are acute but institutional capacity is often limited. In such contexts, dependence on proprietary systems can undermine long-term resilience by tying essential functions—risk mapping, early warning, and service targeting—to external actors.

These challenges are not merely technical. They are institutional and political. Geospatial data are produced, governed, and contested across multiple organizations with overlapping mandates and uneven authority. National mapping agencies, statistical offices, sectoral ministries, space agencies, municipal governments, private firms, international donors, and non-governmental organizations all play roles in shaping what data exist, who controls them, and how they circulate. Funding flows through one set of institutions, while implementation responsibilities rest with another. The authority over base data may be centralized, while analytical capacity is dispersed—or vice versa.

This fragmented institutional landscape helps explain why geospatial projects so often stall at the point of action. Dashboards and portals can satisfy reporting requirements without altering decision processes. Data may be produced at scales or in formats that do not align with local planning needs. Responsibility for maintenance, updating, and adaptation is frequently unclear. Without explicit pathways from spatial evidence to governance—budgeting, regulation, enforcement, or service delivery—maps risk becoming symbolic artifacts rather than instruments of change.

International bodies have increasingly recognized these challenges. The United Nations has formally acknowledged geospatial information as a critical enabler of the Sustainable Development Goals, emphasizing the need for coordinated national geospatial frameworks that integrate policy, legal authority, institutional capacity, and technical infrastructure (United Nations Geospatial Network, 2021; United Nations Committee of Experts on Global Geospatial Information Management, 2025). This work underscores a key point: geospatial value emerges not from data alone, but from how data are embedded in institutions.

For practitioners, the implication is clear. Working with open geospatial data requires more than technical competence. It demands an understanding of provenance, governance, and power. Analysts must ask not only where data come from but also under what conditions they were produced, who maintains them, and who has the authority to act on the insights they generate. These questions are especially important in applied work on environmental risk, where maps can influence investment decisions, land use, and the distribution of harm.

Throughout this book, open data are treated not as a guarantee of equity or effectiveness, but as a starting point. Open datasets and open tools lower barriers to entry, enable replication, and support transparency. They also make it easier for local actors—planners, journalists, community organizations, and public officials—to interrogate claims and participate in spatial debates. But openness alone is insufficient. Durable impact depends on how spatial evidence is translated into governance and action.

The chapters that follow return repeatedly to this tension. Each case illustrates not only how environmental threats can be mapped but also how spatial data move—or fail to move—through institutions. By grounding geospatial analysis in open methods while remaining attentive to institutional realities, practitioners can avoid the trap of "death by dashboard" and use spatial evidence to support decisions that matter.

2.4 The National Geospatial Terrain as a System of Systems

In most Global South countries today, geospatial information does not flow through a single authority or coherent pipeline. Instead, it is produced, governed, and contested across a fragmented institutional landscape that can be understood as a system of systems. This terrain is shaped by overlapping mandates, uneven technical capacity, funding dependencies, and shifting public–private relationships. Figure 2.1 depicts the geospatial institutional terrain for a Global South country and includes national governments, the private sector, international and multilateral organizations, and NGOs and open data communities. Understanding this landscape is essential for anyone seeking to use geospatial data for environmental analysis or policy action.

At the core of this system remain government actors, who continue to provide what is often considered "authoritative" spatial data. National Mapping and Cadastral Agencies (NMCAs) serve as custodians of geodetic frameworks, base maps, and land records, anchoring National Spatial Data Infrastructures (NSDIs). National Statistical Offices increasingly integrate geospatial data into census operations and Sustainable Development Goal

1. National Governments
- National Mapping & Cadastral Agencies
- National Statistical Offices (NSOs)
- Space Agencies

2. Private Sector (Innovation & Infrastructure)
- Software & Platform Giants: Esri (ArcGIS), Hexagon, and Trimble
- Satellite Imagery Providers: Maxar, Planet, Airbus, SatSure, Synspective
- Analytics & AI Startups
- Open Data Foundations: The Overture Maps Foundation

3. International & Multilateral Organizations
- UN-GGIM
- Multilateral Banks: World Bank, International Monetary Fund (IMF), and Inter-American Development Bank (IADB)
- Specialized Agencies: UN-Habitat, FAO, and NASA (SERVIR)

4. NGOs & Open Communities
- OpenStreetMap (OSM)
- Humanitarian Mapping NGOs
- Technical Societies: ISPRS and OGC

FIGURE 2.1
Geospatial players for a typical country. (Source: Based on UN Committee of Experts on Global Geospatial Information Management, 2025.)

monitoring, while emerging national or regional space agencies assert forms of satellite sovereignty that reduce reliance on external imagery providers. Together, these institutions form the regulatory backbone of national geospatial systems.

Surrounding this core is a rapidly expanding private sector, whose role has shifted from that of vendor to strategic partner. Large software and platform firms provide much of the GIS infrastructure used by governments, utilities, and NGOs. Commercial satellite providers dominate access to high-resolution imagery, while specialized analytics and AI firms translate Earth observation data into sector-specific products for agriculture, infrastructure, insurance, and climate risk. At the same time, new "open data foundations" backed by major technology companies are reshaping basemap provision, blurring the boundary between public goods and corporate infrastructure.

A third layer consists of international and multilateral organizations, which exert influence through funding, standards, and capacity-building frameworks. Bodies such as the UN-GGIM articulate global policy directions and promote integrated geospatial frameworks, while development banks finance national modernization projects and data systems. Specialized agencies provide technical assistance in areas such as urban monitoring, food security, and disaster risk reduction. These actors often generate the most detailed spatial data in contexts where domestic capacity is limited, but their involvement can also fragment ownership and accountability.

Finally, NGOs and open communities occupy a distinct and increasingly important position within the geospatial terrain. Volunteer mapping initiatives, humanitarian NGOs, and open technical communities often produce the most granular, locally relevant data—particularly in informal settlements and hazard-prone areas neglected by commercial providers. These efforts expand participation and visibility, but they frequently operate without stable funding or formal authority, limiting their integration into official decision-making processes.

The result is not a hierarchy, but a patchwork. The authority over base data may be centralized, while analytical capacity is outsourced. Funding may flow through international institutions, while implementation rests with under-resourced local agencies. Data may be technically open but institutionally constrained. For practitioners, the challenge is not simply to obtain data, but to understand where it comes from, who controls it, and how it can—or cannot—be mobilized for action.

This book treats this geospatial terrain not as an obstacle to be overcome, but as a reality to be navigated. Effective spatial practice depends less on mastering every system than on recognizing their interactions, limits, and points of leverage.

2.5 GIS as a Way of Working

At its core, GIS is a way of organizing information spatially. It allows relationships that are difficult to see in tables or narratives to become visible when mapped and compared. Elevation influences flood risk, distance shapes access, and boundaries structure governance. GIS makes these relationships explicit.

What distinguishes effective GIS practice from ineffective practice is rarely technical sophistication. More often, it is the clarity of purpose. Analyses that begin with a clear spatial question and a defined decision context tend to remain focused. Analyses that begin with data exploration alone often drift, accumulating maps without producing insight. Throughout this book, GIS is treated as an applied framework for reasoning about place. Maps are not endpoints. They are intermediate representations that help structure arguments, test assumptions, and support decisions.

2.6 Spatial Data Models: Vector and Raster

Nearly all GIS analysis relies on two fundamental data models: vector and raster. Understanding their differences and how they complement one another is essential for environmental mapping. Figure 2.2 depicts raster data in the form of a shaded DEM from the Shuttle Radar Topography Mission on the left and vector data in the form of third-level administrative boundaries from GADM on the right.

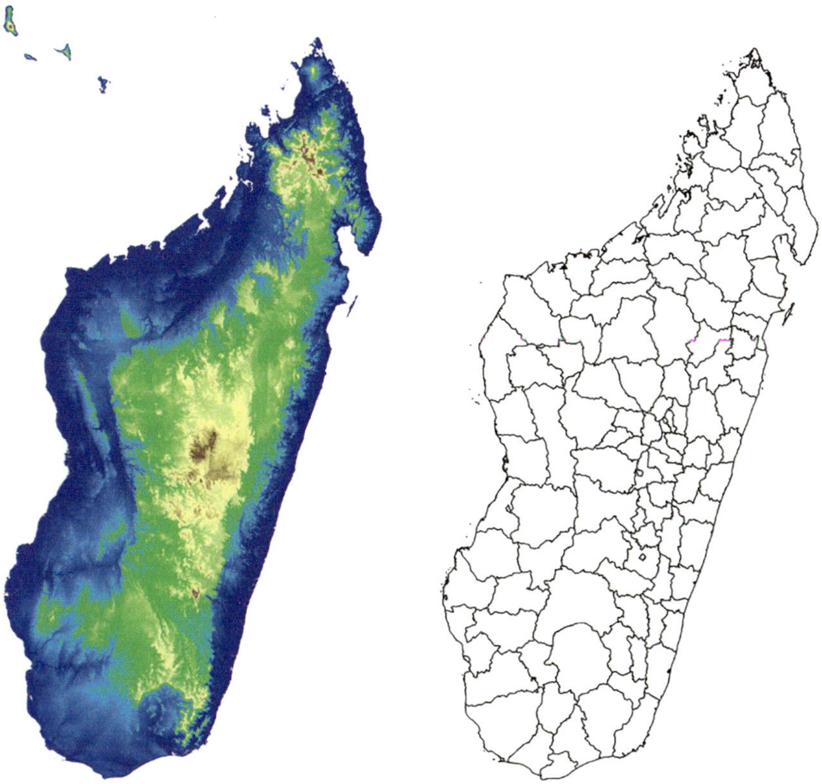

FIGURE 2.2
Raster and vector representations of spatial information. (Source: NASA Shuttle Radar Topography Mission Global. Level 3 administrative boundaries from GADM.)

Raster data represent space as a continuous surface divided into cells, making them well suited for elevation, temperature, and other continuous phenomena. Vector data represent discrete features such as roads, boundaries, and buildings using points, lines, and polygons. Most GIS analyses combine both representations.

Vector data represent discrete features using points, lines, and polygons. Points represent specific locations such as wells, clinics, schools, or air-quality sensors. Lines represent linear features such as roads, rivers, pipelines, or power lines. Polygons represent bounded areas such as neighborhoods, flood zones, land parcels, or administrative districts. Vector features are stored alongside attribute tables that describe their characteristics, including population, land use, service status, or infrastructure capacity. Vector data are particularly well suited to representing governance units, infrastructure networks, and countable features.

A key concept in vector data is topology, which encodes relationships such as adjacency and connectivity. Topology allows GIS to answer questions about which parcels share a boundary, which roads connect to which intersections, or which buildings fall within a floodplain. These relationships are not visually obvious; they are computationally defined.

Raster data divide space into a regular grid of cells, each assigned a numeric value. Rasters are especially well suited for representing continuous phenomena such as elevation, temperature, rainfall, vegetation cover, or pollution concentration. A Digital Elevation Model stores elevation values for every grid cell across a landscape. Satellite imagery, land-surface temperature maps, and vegetation indices are all raster datasets.

Raster resolution, defined by cell size, matters enormously. A 30-m raster captures regional patterns but may miss fine-scale urban variation. A 1-m raster reveals much more detail but requires greater storage, processing power, and interpretive care. Resolution is therefore not a neutral property of data, but a design choice that shapes which risks become visible and which remain obscured.

Most urban hazard analyses combine both data types. Raster layers represent environmental processes, while vector layers represent people, infrastructure, and institutions. GIS becomes powerful precisely because it allows these fundamentally different representations of space to interact.

2.7 Layers and Visual Logic

Most GIS work begins not with analysis but with careful viewing. Spatial data are organized into layers, each representing a distinct set of features or measurements. While this structure may appear technical, it reflects a deeper analytical principle: spatial phenomena are best understood by isolating components and recombining them intentionally.

Each vector layer consists of geometry paired with an attribute table. The attribute table is not ancillary. It is the primary space in which meaning is assigned. A polygon representing a neighborhood is analytically inert until it is associated with attributes such as population, housing age, land use, or service access. In practice, analysts move continuously between map view and table view, testing assumptions and identifying patterns that are not immediately visible spatially (Graser, 2016).

Visual appearance also shapes interpretation. The order in which layers are drawn, the choice of symbols, and the use of transparency influence which features dominate attention. These decisions are not cosmetic. They structure the argument a map makes. A road network drawn boldly over a flood surface suggests accessibility. The same network submerged beneath a semi-transparent raster suggests vulnerability. Good GIS practice therefore treats visualization as an analytical act rather than a final step.

2.8 Core Analytical Operations in GIS

Despite the apparent complexity of GIS software, most environmental hazard mapping relies on a small set of recurring analytical operations. Mastery lies not in tool familiarity but in understanding what these operations mean spatially.

Overlay is foundational. By combining layers, overlay identifies where conditions coincide. A floodplain intersected with housing parcels reveals exposure. Overlay formalizes causal hypotheses about space, making explicit where multiple processes interact—whether environmental, infrastructural, or institutional.

Selection and filtering allow analysts to isolate features based on attributes or spatial relationships. Selecting neighborhoods with low elevation and high population density converts descriptive data into a testable claim about vulnerability.

Distance and proximity analysis operationalize exposure. Buffers translate abstract hazards into measurable zones of influence, converting distance into an operational assumption about exposure. Whether defining populations near highways or facilities reachable within a travel threshold, distance-based operations connect spatial configuration to lived experience.

Raster calculation enables conditional reasoning across continuous surfaces. Raster math allows analysts to define thresholds, combine variables, and model scenarios. Reclassifying elevation into risk categories or combining slope and land cover into a landslide susceptibility surface involves judgment at every step. These classifications are constructed, not discovered, and the assumptions embedded in their thresholds must be documented (Graser, 2016).

Masking constrains analysis to a defined spatial domain. Rather than evaluating a phenomenon everywhere, masking limits calculations to areas inside or outside a specified boundary or condition. In practice, masking is most often applied to raster data, such as restricting elevation or temperature surfaces to an urban area, watershed, or administrative unit. Cells outside the mask are ignored or set to null. Like buffering or reclassification, masking is an analytical decision that defines what is considered relevant to a given question and should be documented when results inform decisions.

Zonal statistics bridge the raster and vector worlds. By summarizing raster values within polygons, analysts translate environmental processes into governance-relevant units. This step is often where scientific analysis becomes legible to policymakers.

Density and heatmapping further abstract spatial patterns. Kernel density surfaces do not represent physical quantities but inferred intensity. Used carefully, they reveal clustering. Used carelessly, they imply precision that does not exist.

Across these operations, the key lesson is consistency. The same analytical logic recurs across hazards. Only the data and stakes change.

2.9 Seeing, Measuring, and Interpreting

A recurring risk in GIS practice is mistaking technical execution for analytical rigor. GIS does not eliminate subjectivity. It redistributes it, shifting judgment from the visible map to the often invisible parameters that produce it. Decisions about classification thresholds, buffer distances, and aggregation units shape outcomes as strongly as data quality.

This book therefore treats GIS not as an answer-producing machine but as a structured way to think spatially. Maps are arguments. Analytical transparency, rather than technical complexity, is what allows those arguments to be challenged, revised, and improved.

2.10 Coordinate Reference Systems and Projections

All spatial data rely on coordinate reference systems to locate features on the Earth's surface. Errors in coordinate systems are among the most common and consequential mistakes in GIS analysis.

A coordinate reference system defines how locations are measured, the units used, and how the curved Earth is represented on a flat map. Geographic coordinate systems such as latitude and longitude are useful for global datasets but poorly suited for distance or area calculations. Projected coordinate systems transform the Earth onto a flat surface, introducing distortion in different ways depending on the projection.

For urban environmental analysis, projections that preserve distance or area are usually preferable. Universal Transverse Mercator (UTM) zones are commonly used for city-scale work because they minimize distortion over limited extents. Good GIS practice requires habitual attention to coordinate systems, including checking the CRS of every dataset, reprojecting layers into a common system before analysis, and confirming that units match intended measurements.

2.11 Scale and Resolution

Environmental threats operate across multiple spatial and temporal scales. GIS analysis must therefore engage scale explicitly.

Spatial scale refers to the geographic extent of analysis. Resolution refers to the level of detail in the data. Temporal scale refers to when data were collected and how frequently they are updated. These dimensions interact. A floodplain map derived from an outdated elevation model may misrepresent current risk. A population dataset updated every decade may fail to capture rapid informal settlement growth. Analysts must continually assess whether data are fit for a particular purpose at a particular moment.

Figure 2.3 displays Mt. Rainier National Park at three different scales to illustrate the design decisions related to data resolution. For the largest-scale

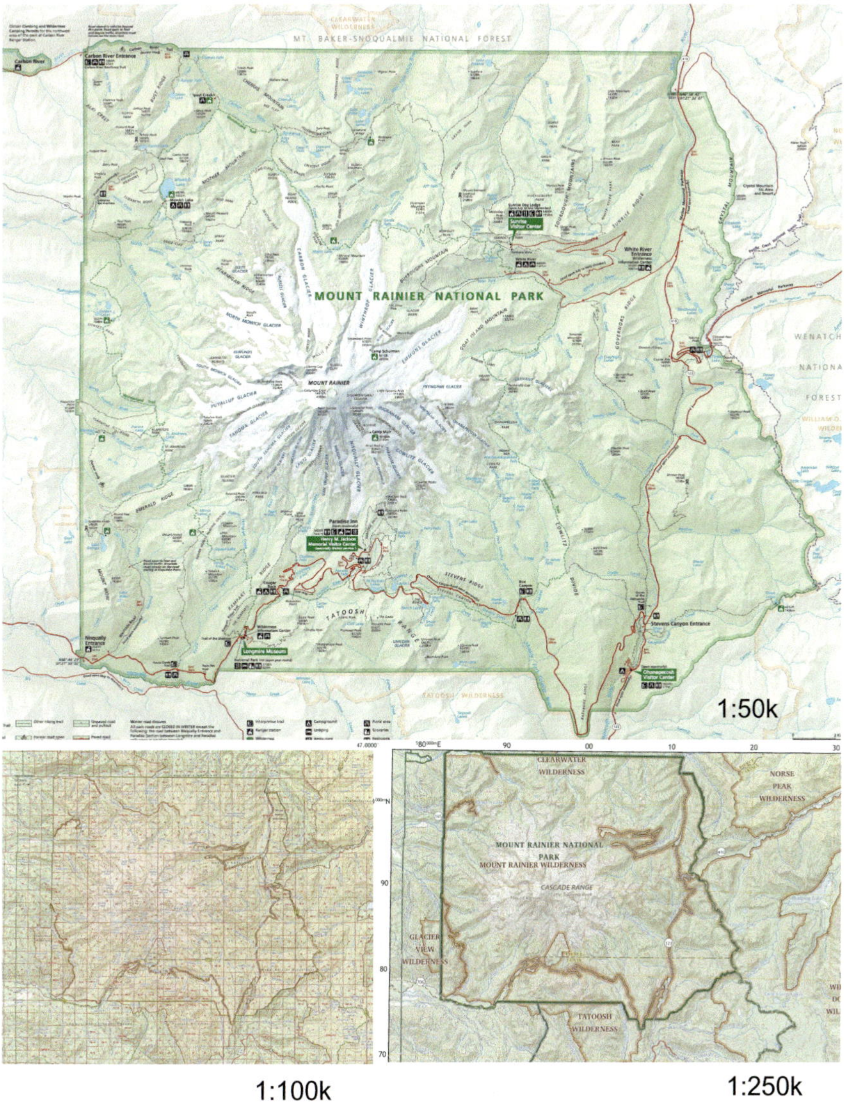

FIGURE 2.3
Scale and resolution in spatial analysis: Mount Rainier is depicted at 1:50,000, 1:100,000, and 1:250,000, revealing countless design decisions. (Source: U.S. Geological Survey. The same National Park appears differently depending on scale and data resolution.)

map, 1:50,000, which is a version of the map handed out to park visitors, amenities such as visitors' centers are labeled. For the 1:100,000-scale map, there is one-half the space for labels and symbolization; for the 250,000-scale map, there is one-fifth the space.

2.12 Metadata and Data Quality

Metadata are not optional. They document who created a dataset, when it was collected, how it was produced, and what its limitations are. In crisis mapping and public decision-making, metadata function as evidence.

Poor metadata can undermine otherwise careful analysis. Well-documented data, even when imperfect, can support transparent and defensible conclusions. GIS does not correct bad data. It amplifies it.

2.13 Open GIS Tools and Foundational Open Data

A practical foundation for GIS work today rests on open-source software and widely available open data. This pairing lowers barriers to entry, supports scrutiny, and reduces dependence on proprietary systems.

QGIS is used as the primary reference environment in this book. It is a mature, actively developed desktop GIS capable of handling vector and raster analysis, cartography, and database connections. Other open tools are often used alongside it, but the analytical logic presented here is not software specific.

Across urban environmental analyses, a small set of foundational datasets appears repeatedly. Elevation data from the Shuttle Radar Topography Mission (SRTM) provide near-global coverage and remain a baseline for terrain analysis. Transportation networks and hydrography are often drawn from OpenStreetMap. Administrative boundaries come from national statistical offices or global repositories. Population data, whether census-based or modeled, provide essential context for exposure analysis. These datasets form the spatial skeleton on which projects are built.

2.14 From Data to Project

Data and software do not constitute a project. GIS work becomes meaningful when it is organized around a question that connects place, process, and decision.

Effective projects begin with a clearly defined spatial question grounded in a specific context. Base data are assembled and checked. Locally collected information is incorporated where possible. Assumptions are tested through simple spatial operations before more complex modeling is attempted. Outputs are produced with the audience in mind.

Locally collected data deserve particular emphasis. Field observations, community mapping, and participatory surveys often reveal conditions absent from official datasets. Even modest local data can change how problems are understood and whose knowledge is recognized.

2.15 Workflow Discipline and Trust

Producing actionable spatial analysis requires a disciplined workflow. Files are organized consistently. Coordinate systems are documented. Data are versioned rather than overwritten. Notes are kept on analytical decisions. These practices allow analyses to be revisited and defended.

Automation can support reproducibility when used carefully. The emphasis should remain on clarity rather than speed. An analysis that can be explained is more valuable than one that merely runs.

2.16 What You Need Now

Practical GIS work in 2026 requires realistic technical capacity. For most urban analyses, a modern computer with adequate memory and storage is sufficient. Reliable internet access supports data acquisition and collaboration, but workflows should function offline once data are downloaded. Time, documentation, and institutional memory are often the most overlooked resources.

Effective GIS work is rarely done in isolation, and local resources matter as much as global ones. Practitioners rely on overlapping communities that provide data, tools, training, and peer support. These networks include open mapping communities such as OpenStreetMap, national statistical offices or census bureaus, other public agencies, educational institutions such as local universities or technical colleges, municipal planning or public works departments, GIS user groups, and online fora and documentation communities.

2.17 Learning about Your Home Place

Before GIS becomes a method, it is often a way of noticing. People learn their home places long before they learn to map them, through repeated encounters and accumulated impressions that rarely announce themselves as knowledge at the time. The slope that always feels steeper than it looks, the street that floods first, the stand of trees that marks a change in soil or wind, and the vacant lot that once held something else. Over time, these observations cohere into an internal map that is partial, uneven, and deeply situated, yet often remarkably precise about how a place behaves.

Such place knowledge does not belong to any single archive or institution. It is carried in practices of looking and remembering, as well as comparing what was expected to what actually happened. It persists in the careful preservation of local records, in the patient observation of landforms and species, in photographs taken for reasons that seemed incidental at the time, and in the stories exchanged among people who have lived with a place long enough to recognize its patterns. This is not informal knowledge in the sense of being careless or unsystematic. It is knowledge produced through duration, familiarity, and attention, shaped by the simple fact of being present across seasons, disruptions, and change.

Digital GIS does not replace this mode of knowing; it sharpens it. What mapping makes possible is the translation of lived recognition into forms that can be tested, shared, revisited, and extended beyond individual memory. Patterns sensed intuitively can be examined explicitly. Suspicions can be checked against evidence. Observations can be layered, compared, and preserved. In this way, GIS becomes not just a technical skill but a means of deepening one's relationship to home. The pleasure that often accompanies this work is not the satisfaction of mastering a tool, but the quieter joy of seeing a familiar place more clearly and discovering that it still has something to teach.

2.18 Looking Ahead

This chapter establishes GIS foundations as a way of working rather than a checklist of tools. The next chapter anchors these concepts in the most consequential dataset for urban environmental analysis: elevation. By beginning with terrain and the idea of home, Chapter 3 grounds abstract GIS principles in the physical landscapes that shape risk everywhere.

References

Chignard, S. (2013). *A brief history of open data*. Paris Tech Review. https://www.paristechreview.com/2013/03/29/brief-history-open-data/

Gayton, I. (2020). *Use of proprietary software in the aid sector perpetuates racial injustice*. https://ivangayton.net/2020/07/03/use-of-proprietary-software-in-the-aid-sector-perpetuates-racial-injustice/

Graser, A. (2016). *Learning QGIS* (3rd ed.). Packt Publishing.
McHarg, I. L. (1969). *Design with nature*. Doubleday.
Marx, R. W., & U.S. Bureau of the Census. (1990). *The Census Bureau's TIGER system*. American Congress on Surveying and Mapping.
Mutinda, R. (2023). *Geographic information systems in developing countries*. Life in GIS. https://www.lifeingis.com/geographic-information-systems-in-developing-countries
United Nations Geospatial Network. (2021). *Geospatial in action: Data and insights for the Sustainable Development Goals*. United Nations.
United Nations Committee of Experts on Global Geospatial Information Management. (2025). *Work plan: Policy and legal framework (2025–2027)*. United Nations. un.org/meetings/GGIM-committee/15th-Session/documents/Workplan_Policy_and_Legal_2025-2027.pdf

3

Starting with Home

Chapter Introduction

Every city begins with its ground. Before infrastructure, before zoning, before hazard maps or dashboards, there is terrain: slopes that shed water, valleys that collect it, and ridgelines that divide neighborhoods and watersheds. The Digital Elevation Model (DEM) captures this underlying structure. Before it becomes data, however, elevation is something people already know through experience: where water lingers after rain, where paths steepen, and where the land itself seems to guide movement and settlement. It is often the first dataset needed for any danger-mapping project because it describes the physical surface upon which all other urban processes unfold (Burrough & McDonnell, 2015; Longley et al., 2015).

This chapter introduces the DEM as a practical entry point into terrain analysis. Rather than treating elevation as an abstract raster layer, it frames the DEM as a way of *reading the landscape*—understanding how gravity, water, and materials move through space, and how those movements shape risk and opportunity in cities. The goal is not technical mastery, but spatial intuition: learning to see why certain places flood, slide, overheat, or remain comparatively safe.

3.1 Why Elevation Matters

Elevation is the hidden architecture behind many urban environmental threats. Floodwaters follow gravity. Landslides initiate on steep slopes. Drainage networks carve pathways that shape informal settlement patterns. Even urban heat islands are influenced by terrain, airflow, and exposure (Montgomery & Dietrich, 1994; Cutter et al., 2003).

Because elevation underlies so many processes, DEMs appear repeatedly throughout this book. They are central to flood modeling, landslide risk assessment, wind exposure analysis, and infrastructure planning (UNDRR, 2015; World Bank, 2021). Starting with elevation allows analysts to build a coherent mental model of place before layering on population, land use, or governance boundaries.

DOI: 10.1201/9781032636948-3

3.2 What is a Digital Elevation Model?

A DEM is a raster dataset in which each cell represents the height of the Earth's surface at that location, usually measured in meters above sea level. Conceptually, a DEM is a numerical description of terrain. Practically, it is a grid of values—often millions of them—stitched together to form a continuous surface (Burrough & McDonnell, 2015).

DEMs can be produced using several methods:

- Radar-based systems, such as NASA's Shuttle Radar Topography Mission (SRTM), which infer elevation from radar reflections (Farr et al., 2007).
- LiDAR (Light Detection and Ranging), which uses laser pulses to measure ground height with very high precision (Maune, 2007).
- Photogrammetry, which derives elevation from overlapping aerial or satellite imagery.
- Drone surveys, increasingly common for local-scale studies.

Each method involves trade-offs between resolution, coverage, cost, and accuracy. No DEM is universally "best." Fitness for purpose matters more than nominal precision (Gesch et al., 2018).

3.3 Finding and Choosing DEM Data

High-quality elevation data are increasingly available through open sources. Commonly used DEM datasets include:

- NASA SRTM (30 m): global coverage, free and widely used (Farr et al., 2007).
- ALOS PALSAR (12.5 m): higher resolution, particularly useful in mountainous terrain (JAXA, 2020).
- Copernicus EU-DEM (25 m): strong coverage for Europe with partial overlap into adjacent regions (Copernicus Programme, 2021).
- National or LiDAR-based DEMs: where available, these provide the highest resolution and accuracy for urban analysis (Maune, 2007).

When selecting a DEM, analysts should consider spatial resolution, vertical accuracy, acquisition date, and whether vegetation and buildings have been removed (bare-earth versus surface models). Using an inappropriate DEM can produce misleading results with unwarranted confidence (Gesch et al., 2018).

3.4 Reading the Landscape Before Modeling it

Before running calculations, analysts should spend time visually interpreting the DEM using hillshade, contours, and color ramps. These representations reveal drainage pathways, ridgelines, and slope breaks, which often correspond directly to observed hazards (Longley et al., 2015).

This visual interpretation builds intuition that later quantitative outputs can be checked against. If a flow-accumulation map contradicts what the terrain clearly suggests, the issue is likely methodological rather than geological (Jenson & Domingue, 1988).

Before downloading data or opening a GIS project, it is worth spending time learning the landscape of one's own community directly. This does not mean conducting formal field surveys or collecting measurements but developing a basic spatial familiarity with how a place works. Walking through neighborhoods, following drainage paths after rain, noticing changes in elevation, and observing where people live, move, and gather all provide forms of spatial information that rarely appear in official datasets.

These observations are not separate from the GIS analysis. They shape how data are interpreted and how projects are designed. A slope that looks insignificant in a raster may feel steep when walked. A drainage channel that appears minor on a map may carry significant flows during storms. Low-lying areas that flood repeatedly may be obvious to residents long before they are visible in a DEM-derived flood model. Paying attention to these cues helps analysts ask better questions of their data.

Getting to know the landscape also involves recognizing social patterns that intersect with terrain. Where housing becomes denser along valley floors, where informal settlements appear on marginal slopes, and where infrastructure follows ridgelines or avoids them all reflect long-standing interactions between people and topography. These patterns provide context for later analysis and help prevent misinterpretation of model outputs.

Importantly, much of this knowledge does not come from what is conventionally labeled as GIS data. Conversations with residents, personal experience, historical memory, and repeated observation over time all contribute to a mental map of place. These forms of knowledge are often uneven, subjective, and difficult to formalize, but they are invaluable for designing analyses that matter.

Starting with the home grounds, geospatial work is embedded in lived experience. It reminds analysts that elevation models and derived products are representations of real landscapes with real consequences. By taking stock of a community before designing a GIS project, analysts create a reference frame against which digital outputs can be evaluated, questioned, and refined.

The value of starting with home becomes clearest when mapping is grounded in movement, conversation, and physical effort, rather than abstract data alone.

3.5 One Perspective on Mapping a Landscape

My First Solo Cartography Project: Discovering Friendship, Resilience, and Beauty in Kenya's Taita Hills, by Primoz Kovacic (2024).

In 2010, I did my first solo cartography project in Africa. Before coming to Kenya in 2009, I had already established a connection with Ronald Mdawida, the director of Kosmos Solutions International, an NGO working in both the Taita Hills and Nairobi's informal settlements. Since I had time to spare while traveling around East Africa, I met Ronnie, and with just a GPS device and a laptop, we headed to the village where he was born, Wongonyi, located near the Kenyan coast in the middle of Taita Hills.

Electricity was still nonexistent within the area at that time. After spending a few days relaxing and getting to know my surroundings, I had a chat with the local chief about my plan to map the village. He decided that I should map the entire Wongonyi Sublocation, covering five villages. The goal was ambitious, to say the least. I planned to chart every home, public structure, and key amenities like water points. I had no idea what to expect but was excited to get started and to see what the final map would reveal. Taita Hills are one big hilly expanse, and for a clearer understanding of the paths and locations of points of interest, I asked my guide, and now my friend, Isaiah, to sketch the locale for me. I later noticed that his sketched map looked very similar to the GPS traces.

Initially, the task seemed manageable, but I quickly felt overwhelmed. The expansive, rugged terrain, under the relentless sun, took me by surprise and showed me just how ill-prepared I was. I suffered blisters on my ears from not wearing a hat and on my feet from walking. Then there was exhaustion from the heat, altitude, and challenging terrain. All of these were challenges to overcome with good planning and preparation, but the kindness of the Wongonyi people proved to be an even bigger obstacle to getting work done in any meaningful time. Each house visit turned into an invitation for tea, meals, or snacks, coupled with conversations about our project, memories, the weather, or life in general. These meetings, while beautiful, were extremely time-consuming. This led Isaiah and me to adopt a stealthier approach to mapping. Avoiding detection, we would emerge from the bushes, record our data, sketch the surroundings quickly, and disappear into the forest, in a style we named "guerrilla mapping."

A month passed, and the map began to materialize before our eyes. The locations of man-made structures were heavily influenced by the terrain, which I wanted to accurately depict, so I created contour lines, essential for the map's final rendition. I tagged all the contour lines with elevation data, so converting the map into a 3D representation was a straightforward step.

This was my first solo experience, far from home, in a country I am yet to know, among the people who would soon become my lifelong friends. This experience also introduced me to the realities of rural life in Kenya. My friends there face daily struggles, from nearly impassable roads to under-resourced public facilities, and scarce employment opportunities.

The experience in Wongonyi was transformative for me personally. It was here where I felt my calling if you will, laying the groundwork for my approach to work and providing valuable lessons for my subsequent work, first at Map Kibera and then at my company Spatial Collective.

3.6 What You Can Do with a DEM

Once prepared, a DEM becomes the foundation for multiple analytical workflows:

- Slope and aspect analysis, relevant for landslides, drainage, solar exposure, and wind.
- Hydrological modeling, including flow direction, accumulation, and watershed delineation (Tarboton, 1997).
- Flood risk assessment, by combining terrain with water levels or rainfall scenarios (World Bank, 2021).
- Infrastructure planning, identifying routes and sites that reduce exposure (UNDRR, 2015).

In practice, DEMs are integrated with vector layers representing roads, buildings, and administrative boundaries. Elevation provides the physical context that gives these layers meaning.

3.7 DEM Processing and Good Practice

Raw DEMs often require preprocessing before analysis. Common steps include:

- verifying coordinate reference systems and vertical units,
- filling sinks and removing artifacts that disrupt hydrological modeling (Jenson & Domingue, 1988),

- clipping to the area of interest, and
- validating elevation values against known reference points or contour maps.

Ground truthing matters. DEMs are models, not reality. Comparing elevation outputs with local knowledge—road grades, riverbanks, and hilltops—helps identify errors before they propagate into downstream analyses (Maune, 2007).

3.8 From Elevation to Flow: Computing Watersheds and Channels

A DEM does more than describe height. When interpreted through the logic of gravity, a DEM becomes a predictive model of movement—especially the movement of water. This is the foundation of watershed and drainage analysis, one of the most powerful and widely used applications of elevation data in environmental risk mapping (Longley et al., 2015).

At its core, watershed analysis rests on a simple assumption: water flows downhill. From each location on a DEM, surface water is assumed to move toward the neighboring cell with the steepest downward slope. When this rule is applied consistently across a raster surface, it produces a flow-direction grid that encodes the direction water would take from every cell under gravity-driven conditions.

Once flow direction is established, analysts can calculate flow accumulation—the number of upstream cells that drain through any given location. Cells with high accumulation values represent areas where water converges. When accumulation exceeds a chosen threshold, these convergent pathways can be interpreted as channels. In this way, drainage networks emerge directly from terrain, without any prior knowledge of mapped rivers or streams.

A watershed is defined as the area upstream of a given point—the land surface from which all surface water drains through that location. Every point on a DEM has an associated watershed, and watersheds therefore nest and overlap. Small catchments feed into larger ones, forming hierarchical drainage systems that reflect the structure of the landscape itself (Longley et al., 2015).

This logic is especially important for understanding flooding. Flood risk is rarely confined to the immediate vicinity of a river channel. Instead, it reflects processes occurring far upstream: rainfall falling on hillsides, runoff moving through informal drainage paths, and accumulation building in low-lying zones. Watershed delineation makes these upstream–downstream connections explicit, revealing how distant terrain contributes to local hazard.

3.9 Thresholds, Channels, and Interpretation

Identifying channels from a DEM requires choosing a flow-accumulation threshold—the minimum number of upstream cells required before flow is considered channelized. This threshold is not a fixed physical constant. It is an analytical choice that depends on terrain roughness, DEM resolution, and the scale of interest. A low threshold produces dense, fine-grained drainage networks; a higher threshold highlights only major channels.

This choice matters for flood analysis. Using too low a threshold can exaggerate drainage density and suggest flood pathways that rarely activate. Using too high a threshold can miss small but consequential channels that concentrate runoff during intense storms—often the very pathways responsible for flash flooding in urban or peri-urban settings.

Longley et al. emphasize that watershed outputs should be interpreted as potential flow pathways, not exact representations of engineered drainage or contemporary river courses. In urban environments especially, channels may be constrained, diverted, or buried. Yet the underlying topographic signal often persists. Even where rivers have been channelized or culverted, floodwaters tend to follow historical drainage routes revealed by the DEM.

3.10 Sinks, Depressions, and Preprocessing

Real landscapes contain closed depressions—lakes, wetlands, and sinkholes—but DEMs also contain artificial depressions created by measurement error or data interpolation. These spurious "sinks" interrupt flow modeling by trapping water unrealistically. For this reason, hydrologic analysis typically begins with a sink-filling step, which raises artificial depressions just enough to allow continuous downslope flow (Jenson & Domingue, 1988; Longley et al., 2015).

This preprocessing step is essential for flood analysis. Without it, modeled watersheds may fragment, channels may terminate abruptly, and accumulation patterns may be misleading. At the same time, analysts must be cautious not to erase meaningful landscape features. The goal is not to impose smoothness, but to restore plausible surface continuity.

3.11 Why Watersheds Matter for Cities

Watershed boundaries rarely align with administrative or planning boundaries. Neighborhoods that appear separate on a map may share the same drainage basin. Conversely, a single administrative unit may straddle multiple watersheds with very different flood dynamics. DEM-based watershed analysis exposes these mismatches, which helps to explain why flooding impacts often cut across jurisdictional lines.

For cities in the Global South, where drainage infrastructure is incomplete or informal, watershed modeling is particularly valuable. It allows analysts to anticipate runoff pathways even when no official drainage maps exist. Combined with rainfall data and land cover, watershed analysis provides a first-order understanding of flood exposure grounded in physics rather than policy.

Later chapters build directly on this logic. Flood modeling, water access analysis, and infrastructure planning all rely—implicitly or explicitly—on the same principles of flow direction, accumulation, and upstream contribution introduced here.

Figure 3.1 depicts the watershed for the McKenzie River in western Oregon outlined with a red line. The map came from mghydro.com, a hydrology online global watershed generator.

Watersheds delineate areas of land that drain toward a common outlet, defined by topography rather than administrative boundaries. Such basin structures shape surface water flow, flood risk, and landscape processes, and provide a foundational framework for elevation-based analysis. Source: Global Watersheds, via Wikimedia Commons.

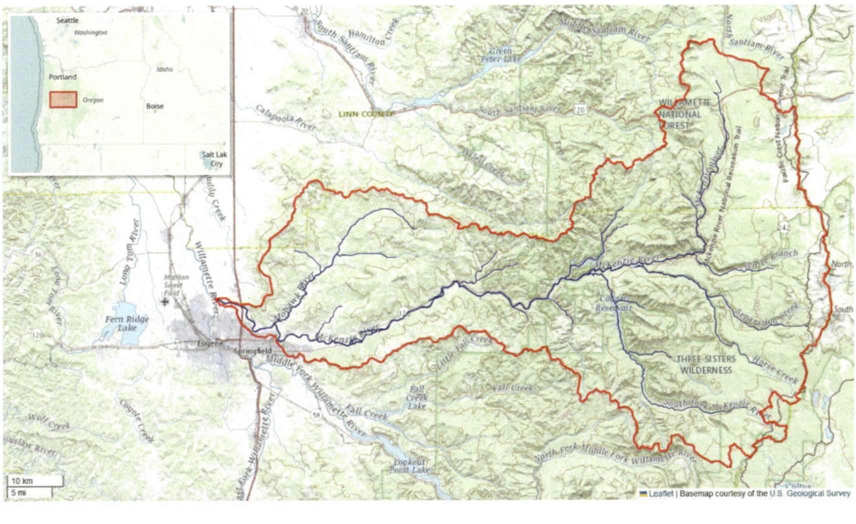

FIGURE 3.1
Map of the McKenzie River Watershed in western Oregon. (Created by Shawn Steinfeger with the free Global Watersheds web app. Permalink to interactive map: https://mghydro.com/watersheds/shared/C6093A.html. Image source: https://commons.wikimedia.org/wiki/Category%3AMaps_of_drainage_basins_from_Global_Watersheds)

3.12 Case Insight: Terrain as Explanation

In cities such as Manizales, Colombia, or Nairobi, Kenya, elevation alone explains much about urban vulnerability. Informal settlements cluster along steep slopes or low-lying flood-prone zones, drainage pathways align with recurrent flood damage, and protected green areas often coincide with terrain unsuitable for construction (Cutter et al., 2003).

A well-chosen DEM visualization can communicate these relationships more clearly than pages of text. Later chapters build on this foundation by layering social and infrastructural data atop terrain.

3.13 Looking Ahead

Elevation shapes the movement of water—both stormwater and drinking water. The next chapter applies the same spatial logic introduced here to model access to essential water resources in cities. Beginning with terrain grounds subsequent analyses in gravity, flow, and constraint.

References

Burrough, P. A., & McDonnell, R. A. (2015). *Principles of geographical information systems* (3rd ed.). Oxford University Press.

Copernicus Programme. (2021). *EU-DEM v1.1 product handbook.* European Environment Agency.

Cutter, S. L., Boruff, B. J., & Shirley, W. L. (2003). Social vulnerability to environmental hazards. *Social Science Quarterly, 84*(2), 242–261. https://doi.org/10.1111/1540-6237.8402002

Farr, T. G., Rosen, P. A., Caro, E., Crippen, R., Duren, R., Hensley, S., ... & Alsdorf, D. (2007). The shuttle radar topography mission. *Reviews of Geophysics, 45*(2). https://doi.org/10.1029/2005RG000183

Gesch, D. B., Oimoen, M. J., Greenlee, S. K., Nelson, C. A., Steuck, M. J., & Tyler, D. J. (2018). The national elevation dataset. *Photogrammetric Engineering & Remote Sensing, 68*(1), 5–11.

JAXA. (2020). *ALOS Global Digital Surface Model (DSM) product description.* Japan Aerospace Exploration Agency.

Jenson, S. K., & Domingue, J. O. (1988). Extracting topographic structure from digital elevation data for geographic information system analysis. *Photogrammetric Engineering & Remote Sensing, 54*(11), 1593–1600.

Kovacic, P. (2024). My first solo cartography project: Discovering friendship, resilience, and beauty in Kenya's Taita Hills. *Medium: Sparks of Humanity*. https://medium.com/sparks-of-humanity/my-first-solo-cartography-project-08fd097a01df

Longley, P. A., Goodchild, M. F., Maguire, D. J., & Rhind, D. W. (2015). *Geographic information science and systems* (4th ed.). Wiley.

Maune, D. F. (Ed.). (2007). *Digital elevation model technologies and applications: The DEM users manual* (2nd ed.). American Society for Photogrammetry and Remote Sensing.

Montgomery, D. R., & Dietrich, W. E. (1994). A physically based model for the topographic control on shallow landsliding. *Water Resources Research, 30*(4), 1153–1171. https://doi.org/10.1029/93WR02979

Tarboton, D. G. (1997). A new method for the determination of flow directions and contributing areas in grid digital elevation models. *Water Resources Research, 33*, 309–319. https://doi.org/10.1029/96WR03137

UNDRR. (2015). *Sendai framework for disaster risk reduction 2015–2030*. United Nations.

World Bank. (2021). *Lifelines: The resilience of infrastructure systems*. World Bank Group.

4

Water Access in the City

Chapter Introduction

Water scarcity is no longer a distant or localized concern; it is a global condition shaping everyday life for billions of people. Recent estimates suggest that roughly one-third of the world's population experiences water scarcity for at least 1 month each year, a figure that continues to rise as climate variability, population growth, and economic development place increasing pressure on finite freshwater resources (UN-Water, 2023; Leijnse et al., 2024). Once framed primarily as a challenge of arid regions or low-income countries, water scarcity is now recognized as a systemic risk affecting cities across the Global South and the Global North alike. From megacities dependent on overstretched reservoirs to smaller urban centers drawing down aquifers faster than they can recharge, the mismatch between water demand and reliable supply has become a defining feature of contemporary urbanization.

At the global scale, this crisis is driven less by absolute shortages of water than by how water is managed, allocated, and governed. A growing body of research combining global hydrological models with empirical case studies shows that water scarcity "hotspots" emerge where hydroclimatic change intersects with rapid urban growth, agricultural and industrial demand, aging infrastructure, and political constraints on reform. These pressures affect both water quantity and water quality, widening the gap between availability and need even in regions not traditionally considered water stressed. As Leijnse and colleagues demonstrate through a comparative analysis of global water scarcity hotspots, the causes and consequences of urban water stress vary widely—but they share common structural roots that make scarcity increasingly persistent and difficult to reverse. Understanding these dynamics requires moving beyond national averages or basin-scale models toward a finer-grained, geographically explicit view of how water access actually functions within cities. This chapter approaches urban water scarcity as a spatial problem, showing how geospatial data and mapping tools can reveal the uneven landscapes of access, risk, and governance that underlie contemporary water crises.

DOI: 10.1201/9781032636948-4

4.1 Setting the Scene

In early 2018, residents of Cape Town confronted a scenario that had previously seemed unthinkable for a major global city: the possibility that municipal taps would be turned off entirely. Years of below-average rainfall, combined with rapid population growth and long-standing constraints on supply infrastructure, pushed reservoir levels to historic lows. City officials announced a looming "Day Zero," the point at which household water service would cease, and residents would be required to collect a strictly rationed allotment from centralized distribution points. Images of empty dams, household water meters, and residents queuing at communal taps circulated worldwide, transforming Cape Town into a symbol of urban water vulnerability in an era of climatic uncertainty (Pascale et al., 2020; Brookings Institution, 2019).

"Day Zero" quickly entered the global policy lexicon because it named a threshold rather than a drought: a single moment when social order, public health, and economic activity would all be reshaped by the absence of reliable water. What made the crisis especially instructive was not only its severity but also its geography. Water shortages did not affect all neighborhoods equally. Affluent areas often had storage tanks, private boreholes, and the financial means to purchase water, while poorer communities faced stricter rationing and longer travel distances. The prospect of Day Zero made visible how water access is structured by infrastructure, elevation, income, and governance, not simply by rainfall.

Cape Town ultimately avoided Day Zero through a combination of aggressive demand management, public communication campaigns, and behavioral change. Daily per capita water use was cut by more than half, emergency regulations restricted nonessential consumption, and residents were enlisted as active participants in conservation efforts. Analysis by the Brookings Institution (2019) emphasized that the city did not solve its crisis by discovering new water sources, but by reorganizing how water was valued, monitored, and governed under extreme scarcity. The episode demonstrated both the fragility and the adaptability of urban water systems when faced with acute stress.

Far from being an isolated event, Cape Town's experience now appears as an early warning. In Iran, more currently, a prolonged combination of drought, over-extraction, and infrastructure mismanagement has pushed several major cities toward their own version of Day Zero. In 2024 and 2025, officials in the cities of Tehran, Karaj, and Mashhad—together home to more than 16 million people—warned that reservoir levels had fallen to critically low levels, in some cases below 10 percent of capacity. Reporting from the BBC documented how decades of deferred investment and politically constrained water governance compounded hydroclimatic stress, raising the prospect of widespread service interruptions in one of the Middle East's largest urban regions (BBC reporting; UN-Water, 2023).

Similar dynamics have played out elsewhere over the past decade. In India, Chennai's reservoirs ran dry in 2019 after years of groundwater depletion and erratic monsoon rains. Southern European cities, including those in Spain and Italy, now contend regularly with drought restrictions once associated primarily with arid regions. In São Paulo, Brazil, a city often perceived as water-rich, system failures during the mid-2010s exposed how quickly urban supply can be disrupted even when average rainfall appears sufficient. Across these cases, the pattern is consistent: water crises emerge where climatic variability intersects with governance failures, aging infrastructure, and unequal access (Gleick, 2018; UN-Water, 2023).

Taken together, these events underscore a central lesson of urban water scarcity. Shortages are rarely caused by a lack of water alone. Instead, they reflect how water systems are designed and managed over time. Crises such as Day Zero reveal the hidden geographies of access embedded in cities—geographies that GIS and geospatial tools in general are uniquely suited to expose. Understanding these patterns is a necessary first step toward addressing the deeper forces now stressing urban water systems worldwide.

4.2 Understanding the Threat

4.2.1 Threat Overview: What is Stressing Urban Water Systems?

Urban water systems around the world are under growing strain (UN-Water, 2023). Global freshwater withdrawals increased by approximately a factor of six over the twentieth century, reflecting population growth, agricultural expansion, and industrialization (Gleick 2018). Cities now compete with agriculture and industry for limited supplies, while aging infrastructure and climate variability reduce reliability. These pressures are intensifying fastest in rapidly urbanizing regions but are increasingly evident in high-income countries as well.

At its core, the contemporary urban water crisis reflects a widening gap between water demand and the systems designed to meet it. Cities concentrate people, economic activity, and infrastructure, making them uniquely vulnerable to disruptions in supply. While climate change has altered precipitation patterns and increased the frequency of droughts in many regions, hydroclimatic stress alone does not explain the scale or persistence of recent water shortages. Instead, urban water systems are being stressed by a combination of demographic, infrastructural, and institutional forces that compound one another over time.

Rapid urban population growth is one of the most important drivers. Cities in the Global South are absorbing millions of new residents through natural increase and migration, often extending far beyond the capacity of existing water networks. Pipes laid decades ago often under colonial administrations were never designed to serve today's populations or urban footprints. As demand rises, utilities frequently rely on the same reservoirs, rivers, or aquifers, pushing systems closer to their physical limits. Over-extraction of groundwater, in particular, has become a defining feature of urban water stress, lowering water tables, increasing pumping costs, and degrading water quality through salinization and contamination.

Infrastructure age and maintenance represent a second, closely related threat. In many cities, water losses from leaking pipes exceed 30 percent of total supply, meaning that a substantial share of treated water never reaches users (Kingdom et al. 2006). Deferred maintenance, limited revenue, and political resistance to rate increases leave utilities unable to invest in upgrades. The result is a cycle of declining service reliability marked by intermittent supply, low pressure, frequent main breaks, and delayed repairs. These technical failures disproportionately affect low-income neighborhoods located at the ends of distribution networks or at lower elevations, where pressure is weakest and service interruptions are most common.

Climate variability and climate change intensify these existing vulnerabilities. Prolonged droughts reduce surface water availability and limit aquifer recharge, while extreme rainfall events can overwhelm treatment facilities and contaminate sources. It's important to never lose sight of the fact that climate impacts are filtered through existing governance systems. Cities with flexible management, transparent communication, and diversified supply portfolios are better able to adapt, while those with fragmented authority or politicized water management face higher risks of crisis. As recent studies of global water scarcity hotspots emphasize, the same climatic stress can produce very different outcomes depending on how water systems are managed and regulated.

Finally, urban water stress must be understood as a problem of allocation as much as supply. Industrial users, irrigated agriculture, and energy production often consume large volumes of water, sometimes at subsidized rates, even as households face rationing or rising prices. These trade-offs are embedded in legal frameworks and political economies that shape who bears the costs of scarcity. In this sense, urban water shortages are not simply technical failures or natural disasters; they are the cumulative outcome of choices about infrastructure and governance made over decades.

4.2.2 Unpacking Drivers of Urban Water Scarcity

To understand why urban water shortages are becoming more frequent and persistent, let's go beyond single-cause explanations and instead examine how multiple forces interact within coupled human–water systems.

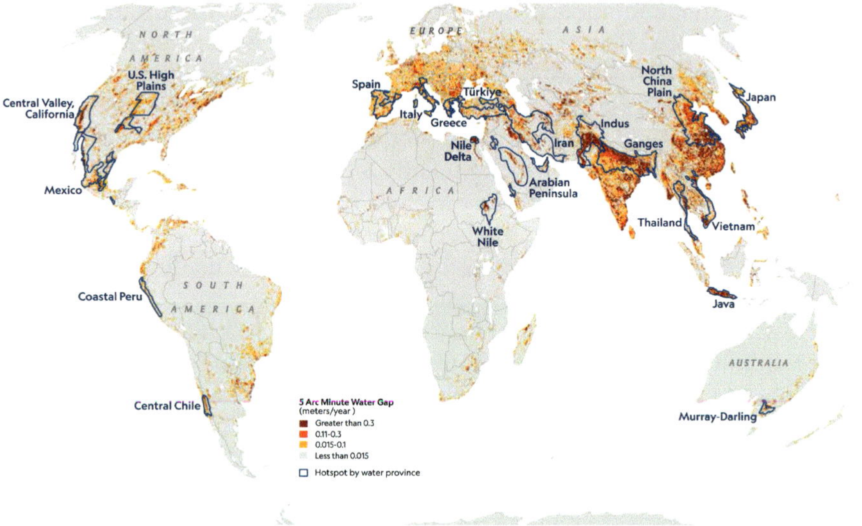

FIGURE 4.1
Global water scarcity hotspots and population exposure. (Source: Leijnse et al. (2024), *Environmental Research Letters*; adapted from PCR-GLOBWB 2 global hydrological model output. Article is open access.)

One widely used analytical approach is the Drivers–Pressures–State–Impacts–Responses (DPSIR) framework, which provides a structured way to trace how social and environmental processes combine to produce water scarcity outcomes. Recent global research applying DPSIR to water scarcity hotspots highlights the complexity and diversity of pathways through which cities experience water stress (Leijnse et al., 2024).

Approximately one-third of the global population experiences water scarcity for at least 1 month per year. As shown in Figure 4.1, hotspots emerge where hydroclimatic variability intersects with population growth, agricultural and industrial demand, and infrastructure constraints. The world map depicts water scarcity hotspots around the world, with stress regions stretching from California's Central Valley to India's Ganges river system to Japan.

Within this framework, drivers include long-term forces such as population growth, economic development, urbanization, and climate change. These drivers generate pressures on water systems, including increased withdrawals for municipal, agricultural, and industrial use, groundwater over-extraction, pollution, and altered hydrological regimes. Together, these pressures change the state of water systems by reducing available supply, degrading water quality, and increasing variability in flows and storage. The resulting impacts are felt unevenly across cities, manifesting as intermittent service, rising costs, public health risks, and heightened vulnerability for low-income and informal settlements. Finally, responses—ranging from infrastructure investments and conservation policies to emergency rationing and governance reforms—attempt to mitigate or adapt to these impacts, with varying degrees of success.

A key insight from global DPSIR analyses is that water scarcity hotspots rarely share a single dominant cause. Instead, they cluster around common combinations of drivers and pressures, producing distinct patterns of risk and response. For cities, this means that effective solutions depend not only on hydrological conditions but also on how urban water systems are governed, financed, and integrated into broader development strategies.

4.2.3 Global Water Scarcity Hotspots: Patterns, Clusters, and Implications

Global analyses of water scarcity reveal that urban water stress is neither random nor evenly distributed. Figure 4.1 shows the 21 major water scarcity hotspots identified by Leijnse et al. (2024), based on modeled water gaps between demand and availability during the period 2010–2019. These hotspots are defined at the water-province scale and represent regions where withdrawals consistently exceed the sustainable supply. While many hotspots coincide with arid and semi-arid climates, others appear in regions with moderate rainfall, underscoring that scarcity is shaped as much by human use and governance as by physical hydrology.

A key contribution of the Leijnse et al. study is its identification of seven distinct clusters of water scarcity hotspots, each characterized by different combinations of drivers, pressures, impacts, and responses. Some clusters are dominated by rapid population growth and urban expansion, while others by intensive irrigated agriculture, industrial withdrawals, or governance constraints that limit adaptive capacity. This clustering helps explain why similar climatic conditions can produce very different urban outcomes. Cities facing comparable drought conditions may experience sharply contrasting levels of service reliability depending on institutional strength, infrastructure investment, and demand management practices.

Importantly, the hotspot map challenges the assumption that water scarcity is confined to the Global South. Southern Europe, parts of the western United States, the Middle East, South Asia, and northern China all emerge as areas of persistent stress. In these regions, growing urban demand competes directly with agriculture and industry, while climate variability amplifies seasonal and interannual shortages. The map also highlights how cities often sit downstream—literally and politically—of decisions made elsewhere in a basin, making them vulnerable to overuse beyond their administrative boundaries.

The situation that has recently unfolded in Iran illustrates how multiple DPSIR pathways can converge. In Tehran, Karaj, and Mashhad, decades of population growth and groundwater over-extraction have coincided with declining reservoir storage and limited infrastructure renewal. These pressures have altered the state of the system, creating lower storage, reduced redundancy, and increasing dependence on stressed sources. Impacts now threaten continuous urban water service for more than 16 million residents. While hydroclimatic change plays a role, the crisis reflects long-standing governance challenges that constrain timely responses.

Similar dynamics are visible in parts of Europe, where drought and water scarcity have become recurrent policy concerns rather than exceptional events. Spain and other Mediterranean countries now regularly impose restrictions on urban and agricultural water use, tracked using indicators such as the Water Exploitation Index Plus (WEI+). These cases demonstrate that even well-resourced regions are vulnerable when demand approaches or exceeds sustainable supply. Together, global hotspot patterns and regional clusters underscore a central lesson for cities: water scarcity is not a singular threat, but a family of related risks that require context-specific, spatially informed responses.

4.2.4 Endemic System Weaknesses: Infrastructure and Governance

Urban water scarcity is rarely caused by a sudden system collapse. More often, it emerges from long-standing weaknesses embedded in the physical and institutional architecture of water supply networks (Gleick, 2018; UN-Water, 2023). Pipes, pumps, treatment plants, and billing systems are expected to deliver a continuous public service, yet in many cities, these systems operate far beyond their intended design life. The result is not total failure, but chronic unreliability—intermittent flow, low pressure, frequent breakdowns, and uneven service that disproportionately affects the most vulnerable residents (OECD, 2016; Kumpel & Nelson, 2016).

Aging and inadequate infrastructure is a central vulnerability. In many cities in the Global South, water networks were laid during colonial periods or early post-independence expansion, designed to serve smaller populations and narrower urban footprints (Nilsson & Nyanchaga, 2008; Foster & Briceño-Garmendia, 2010). Decades of rapid growth have extended cities outward without commensurate investment in trunk lines, storage capacity, or treatment facilities. Leakage rates exceeding 30 percent are common, meaning that large volumes of treated water never reach users (Kingdom et al., 2006). Where pressure fluctuates, pipes are more likely to fail, contamination risks increase, and utilities are forced into reactive repair cycles rather than planned maintenance (Kumpel & Nelson, 2016).

These technical weaknesses are compounded by governance constraints. Water utilities often operate under political pressure to keep rates low, even as costs rise and infrastructure deteriorates (Whittington et al., 2009). Limited revenue undermines the ability to invest in repairs, staff training, and monitoring technologies. In some contexts, fragmented authority—split among municipal agencies, regional water boards, and national ministries—creates unclear accountability when failures occur (Batley & McLoughlin, 2015). Where corruption or informal political influence shapes access to connections or enforcement, system performance declines further (Davis, 2004). The cumulative effect is a water network that functions unevenly, privileging some users while marginalizing others.

Intermittent supply is a particularly revealing symptom of system stress. When water flows only a few hours per day or a few days per week, households with resources adapt by installing rooftop tanks, pumps, and private boreholes. Those without such means must queue at community taps, purchase water from vendors at inflated prices, or carry water long distances (Thompson et al., 2001; Graham et al., 2016). In this way, infrastructure weakness translates directly into social inequality. Service interruptions do not simply reflect scarcity; they actively reproduce disparities by shifting costs—time, money, and health—onto the poorest residents (Peloso & Morinville, 2014).

Informal water systems often emerge to fill these gaps. Vendors, tanker trucks, standpipes, and kiosks provide essential services where municipal networks fall short, but they also introduce new vulnerabilities. Prices are typically higher, quality is harder to regulate, and supply can be unpredictable (Whittington et al., 1991; Komives et al., 2005). While informal provision demonstrates local resilience and adaptability, it also signals systemic failure: a parallel water economy operates because the formal system cannot meet universal needs (Allen et al., 2006).

Taken together, these endemic weaknesses explain why urban water crises tend to follow familiar patterns. Infrastructure degrades, governance struggles to keep pace with growth, and inequalities widen as scarcity intensifies. These conditions do not arise overnight, nor are they evenly distributed across the city. They are spatially patterned outcomes of historical investment decisions, political priorities, and infrastructural neglect—patterns that geospatial analysis is uniquely well suited to reveal (Graham & Marvin, 2001; McFarlane, 2008).

In Figure 4.2, children gather near a community standpipe in an African setting, some holding brightly colored plastic tubs. In informal systems, women and children are often burdened with the task of bringing home water for household use.

4.2.5 Informal Systems and Adaptive Workarounds

Where formal water systems fail to deliver reliable service, informal and semi-formal arrangements emerge to fill the gap. Across cities in the Global South, water access is often maintained through a patchwork of community taps, kiosks, tanker trucks, boreholes, and private vendors operating alongside—or entirely outside—municipal networks. These systems are not marginal anomalies; they are central components of how urban water actually functions under conditions of scarcity and institutional constraint.

Informal water provision reflects both necessity and ingenuity. In neighborhoods with intermittent supply, residents may coordinate shared storage tanks and install private pumps to draw water during low-pressure hours.

FIGURE 4.2
Endemic weaknesses in urban water systems: Children carrying water at a community tap. safariconsoler_communitytap.jpg. (Photo by Safari Consoler: https://www.pexels.com/photo/children-holding-bowls-and-plates-11738251/ CC: free to use.)

Or they might use vendors who transport water from distant sources. Community-managed standpipes and kiosks, sometimes supported by NGOs or donor-funded programs, provide another layer of access, particularly in informal settlements excluded from formal service areas. These arrangements demonstrate local adaptive capacity, allowing households to cope with unreliable infrastructure in the absence of universal municipal provision.

At the same time, informal systems introduce new vulnerabilities. Prices for vendor-supplied water are often several times higher than municipal tariffs, imposing disproportionate financial burdens on low-income households. Quality control is uneven, as water may be sourced from unregulated wells or transported in unsanitary containers. Supply can be unpredictable, particularly during droughts or infrastructure failures, when competition among users intensifies. In some cases, informal providers operate in legally ambiguous or outright illicit conditions, complicating efforts to regulate pricing, safety, or equity.

Importantly, informal systems do not merely compensate for technical shortcomings; they reflect governance choices embedded in urban water management. Exclusion from formal networks is often a function of land tenure status, political marginalization, or historical planning decisions rather than engineering feasibility alone. Informal provision thus becomes normalized, masking systemic inequities while allowing utilities and governments to defer investment in network expansion. Over time, this dual system—formal for some, informal for others—solidifies into a durable spatial pattern of unequal access.

From a geospatial perspective, informal water systems are difficult to map precisely, yet their spatial logic is clear. They cluster in peripheral neighborhoods, low-lying areas with weak pressure, and informal settlements beyond official service boundaries. Understanding where and how these systems operate is essential for any serious attempt to model urban water access, as they shape both the availability and the lived experience of water in ways that conventional utility maps often fail to capture.

4.2.6 Human Impacts: Time, Cost, and Risk

The consequences of unreliable water access are borne most directly at the household level. When taps do not run consistently, the burden of securing water shifts from centralized infrastructure to individuals, transforming water access into a form of daily labor (Thompson et al., 2001; WHO & UNICEF, 2021). Time spent queuing at community taps, walking to distant sources, or waiting for tanker deliveries represents a hidden cost rarely, if ever, captured in official statistics. These time burdens fall disproportionately on women and children, reinforcing gendered patterns of unpaid labor and limiting opportunities for education and income generation (Sorenson et al., 2011; Graham et al., 2016).

Financial costs compound these pressures. Households reliant on vendors or kiosks often pay far more per liter than those connected to municipal networks, despite having lower and less stable incomes (Whittington et al., 1991; Komives et al., 2005). During periods of acute scarcity, prices can spike dramatically, forcing families to ration consumption, reduce spending on food or healthcare, or rely on unsafe sources. In this way, water scarcity acts as a regressive force, deepening existing urban inequalities rather than distributing costs evenly across the population (UN-Water, 2023).

Health risks further amplify the human impact of informal and intermittent water access. Intermittent supply increases the likelihood of contamination, as pressure drops allow pathogens to enter pipes and stored water degrades over time (Kumpel & Nelson, 2016). Water sourced from unregulated wells, surface sources, or reused containers poses additional risks, particularly for children, who are more vulnerable to waterborne disease (Fewtrell et al., 2005). In many cities, outbreaks of diarrheal illness and other water-related diseases are linked not to absolute shortages, but to instability and fragmentation within supply systems under stress (WHO, 2017).

These impacts are not evenly distributed across the city. Within the same metropolitan area, some neighborhoods experience continuous service, while others receive water only a few hours per week. Such disparities are rarely accidental. They reflect the cumulative effects of infrastructure layout, elevation, political influence, land tenure, and historical investment patterns (Peloso & Morinville, 2014; Graham & Desai, 2022). Mapping these differences makes visible the social geography of water scarcity, revealing who absorbs the costs of system failure and who is shielded from them.

Together, these human impacts underscore why urban water scarcity cannot be understood solely in technical terms. Access to water shapes daily routines, household finances, health outcomes, and social relations. By tracing how systemic weaknesses translate into lived experience, this chapter sets the stage for spatial analysis that connects infrastructure, governance, and human well-being—an essential step toward designing more equitable urban water systems.

CASE BOX: BLANTYRE, MALAWI—WALKING FOR WATER IN A SINGLE CITY

Blantyre, Malawi's second-largest city, illustrates how water access can vary dramatically within a single urban area, even among neighborhoods that appear similar in income or housing type. Research led by Caetie Ofiesh, conducted in collaboration with local partners and Water for People, documented water access using detailed household surveys combined with walking distance and time measurements rather than relying solely on utility coverage maps (Ofiesh, 2011; Water for People, 2021). This approach, as shown in Figure 4.3, revealed a fragmented landscape of access shaped by reliability, distance, and time, rather than by the simple presence or absence of infrastructure.

In several informal settlements, residents technically lived "within service areas," yet taps flowed only intermittently—sometimes for just a few hours per week. During outages, households relied on shared standpipes, shallow wells, or surface water sources, often walking significant distances multiple times per day (Ofiesh, 2011). Even small variations in elevation, pipe pressure, or proximity to functioning infrastructure produced stark differences in daily water burdens between adjacent neighborhoods. These microgeographies of access would remain invisible without spatially explicit, ground-level data.

The Blantyre study highlights two critical insights for urban water mapping. First, distance alone is an incomplete measure of access. A nearby water point may be unusable if supply is unreliable, queues are long, or costs fluctuate (WHO & UNICEF, 2021). Second, citywide averages obscure localized hardship. Within the same municipality, households may experience entirely different water realities depending on where they sit within the distribution network. By integrating household surveys with geospatial analysis, the Blantyre work demonstrates how large-N, fine-scale data collection can reveal inequities embedded within urban water systems—inequities that conventional utility maps often fail to capture.

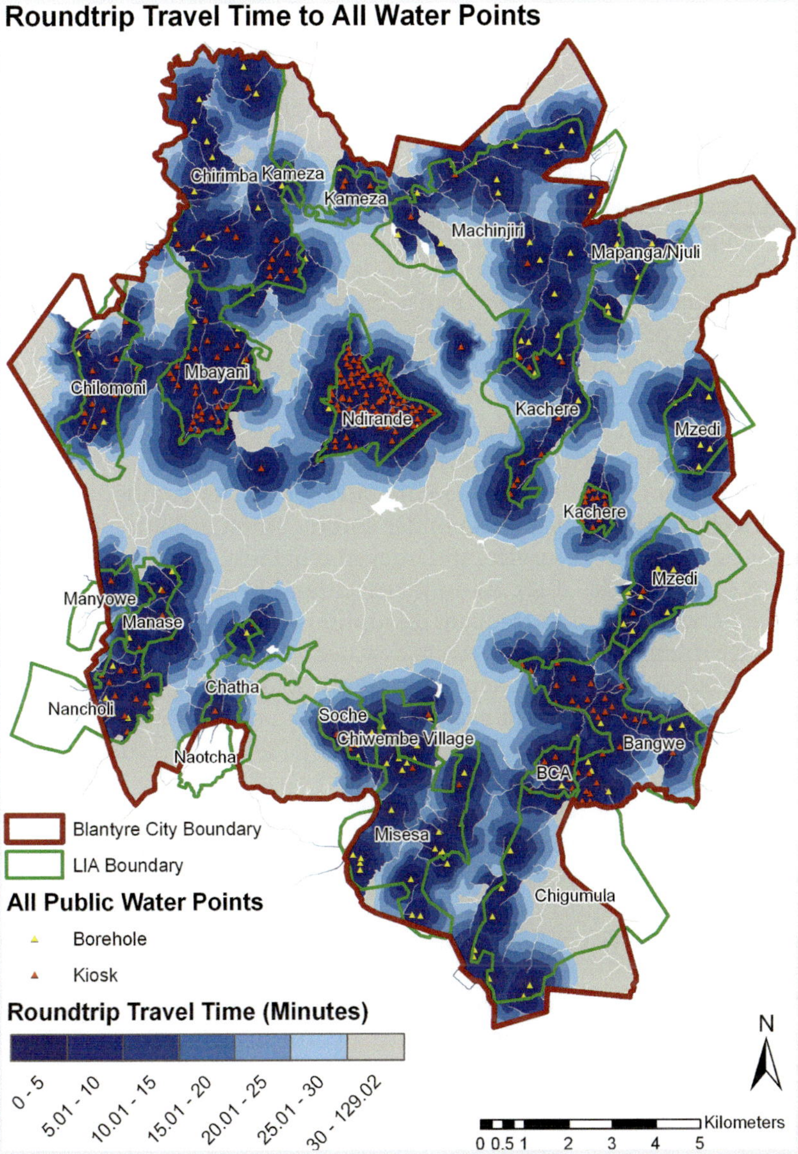

FIGURE 4.3
Blantyre Water Access Map for informal communities. 2,550×3,300 pixels. (Reproduced with permission.)

4.2.6.1 One Affected Place: Accra, Ghana—Overdraft, Growth, and the Limits of Supply

Accra presents a different but equally instructive case of urban water stress, driven by rapid metropolitan growth layered onto finite supply systems. Scholars of Accra's urban transformation note that rapid metropolitan expansion has consistently outpaced infrastructure investment, producing uneven and unreliable access to basic services—including water—particularly in low-income and peri-urban areas (Grant, 2009). Much of the city's water infrastructure was established during the colonial period and expanded incrementally after independence, but it was never designed to support the scale or spatial extent of today's Greater Accra Metropolitan Area (Songsore, 2008; Gough & Yankson, 2012). As in-migration from rural areas and smaller cities has expanded the urban footprint, demand for water has risen far faster than investment in new supply or distribution capacity.

Accra relies primarily on surface water drawn from the Volta River system, through use of the artificial Lake Volta and Akosombo Dam, supplemented by smaller sources and limited groundwater extraction (Ghana Water Company Limited, 2018). Over time, withdrawals have approached—and in some periods exceeded—reliable yields, particularly during dry seasons. Population growth, industrial demand, and losses from aging distribution networks have placed the system under chronic strain, resulting in rationing schedules, intermittent service, and uneven reliability across neighborhoods (Peloso & Morinville, 2014). As in many cities, low-income and peripheral areas experience the most frequent disruptions.

One distinctive feature of Accra's water economy is the widespread use of packaged drinking water, commonly sold in plastic sachets. While often framed as evidence of system failure or privatization, sachet water occupies a more ambiguous role. Research by Justin Stoler and colleagues has shown that sachets can improve short-term access to microbiologically safer drinking water—particularly for children—even as they raise concerns about affordability, plastic waste, and long-term sustainability (Stoler et al., 2012; Stoler et al., 2015). The sachet economy thus reflects both adaptation and inequity: a market solution that fills gaps left by an overstretched public system while shifting costs onto households least able to absorb them.

Accra's experience underscores a central theme of urban water scarcity: growth without commensurate infrastructure investment produces systemic overdraft. The resulting shortages are not simply hydrological failures, but outcomes of political, economic, and planning decisions made over decades (Peloso & Morinville, 2014; UN-Water, 2023). Mapping these dynamics—sources, flows, rationing zones, and informal alternatives—reveals how urban water stress is spatially structured and socially uneven, reinforcing the value of geospatial approaches that link water supply and lived experience.

4.2.7 Water Supply Fundamentals: Sources, Recharge, and System Limits

Urban water systems draw from two primary sources: surface water and groundwater. While these sources are often discussed separately, they function as interconnected components of a single hydrological system shaped by climate, geology, land use, and human management (Gleick, 2018; UNESCO, 2023). Understanding the basic mechanics of how cities obtain and sustain water supply is essential for interpreting patterns of scarcity and risk.

Surface water includes rivers, lakes, reservoirs, and impoundments that capture runoff from surrounding watersheds. Many large cities depend on surface water systems regulated by dams and storage reservoirs, which buffer seasonal variability and support continuous supply. However, surface water availability is highly sensitive to climate variability and change. Prolonged drought reduces inflows, while extreme precipitation can overwhelm treatment facilities and degrade water quality through sedimentation and contamination (IPCC, 2022). Surface water systems are also vulnerable to upstream withdrawals, land-use change, and basin-scale governance decisions, placing cities at risk from actions taken far beyond their administrative boundaries (Grafton et al., 2018).

Figure 4.4 depicts the steps in the water provision process from the source (surface water, groundwater, and reuse water) through transmission, utility reservoir, water treatment, distribution, and withdrawal for industry, residential, and commercial.

Groundwater is stored in subsurface aquifers and accessed through wells and boreholes. In many cities—particularly in arid and semi-arid regions—groundwater serves as a critical supplement or backup to surface supply, and in some cases, the dominant source of urban water (Foster and Chilton 2003). Aquifers recharge through the slow infiltration of rainfall and surface water, a process that can take months to centuries depending on geology and soil conditions. When withdrawals exceed recharge, groundwater levels decline, increasing pumping costs and reducing long-term availability. In coastal

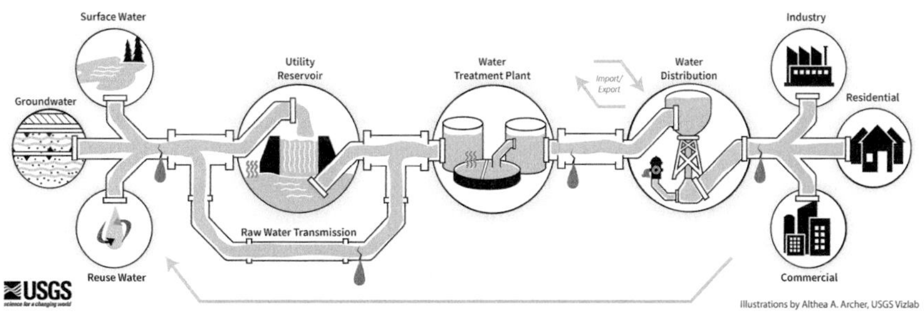

FIGURE 4.4
Typical water supply system in the United States. (Credit: Althea Archer and Richard Niswonger (USGS). Public Domain.)

cities, sustained over-extraction can lead to saltwater intrusion, permanently degrading freshwater reserves (Konikow and Kendy 2005).

Urban well fields operate by distributing pumping across multiple wells to minimize drawdown and maintain system pressure. Under conditions of rapid demand growth, utilities may deepen wells or increase pumping rates without fully accounting for recharge constraints. Such strategies can temporarily stabilize supply while masking structural deficits that only become visible during droughts or infrastructure failures (Wada et al., 2014). Once groundwater systems are depleted or contaminated, recovery is slow and uncertain, making overdraft one of the most persistent and difficult forms of urban water stress.

Recharge dynamics are further shaped by urban form. Impervious surfaces reduce infiltration, while stormwater drainage systems rapidly convey rainfall away from recharge zones. Seasonal rainfall variability compounds these effects, creating short periods of apparent abundance followed by extended scarcity (Shiklomanov & Rodda, 2003). When supply systems fail—through drought, infrastructure breakdown, or governance disruption—cities often turn to emergency measures such as water trucking, temporary boreholes, or interbasin transfers, which are costly and rarely sustainable.

These supply fundamentals underscore why urban water scarcity is not simply a problem of finding new sources. Surface and groundwater systems operate within physical limits that cannot be bypassed indefinitely. Effective water management requires balancing withdrawals with recharge, diversifying sources, and anticipating system stress before crisis emerges. For geospatial analysis, this means mapping not only where water is consumed but also where it originates, how it moves through the system, and how resilient those pathways are under increasing pressure.

4.2.8 Global Data Gaps and Measurement Challenges

A persistent obstacle in mapping urban water access is that the most policy-relevant questions—*who has reliable water, at what cost, of what quality, and how often*—are rarely captured in consistent, spatially explicit datasets. Global reporting frameworks have made major progress in tracking access to "improved" or treated drinking water, but these indicators are typically compiled at national or broad subnational scales and often cannot resolve the intra-urban inequalities that define water scarcity in cities (WHO & UNICEF, 2021; UN-Water, 2023). Even when utilities publish service-area maps, they may describe network extent rather than actual delivery reliability, pressure, continuity, or affordability—variables that often matter more to households than nominal connection status (Kumpel & Nelson, 2016).

Several structural factors drive these data gaps. First, utilities often lack complete, up-to-date asset and customer records, especially in rapidly growing cities where informal expansion outpaces formal connection processes. Network maps may be incomplete, analog, proprietary, or politically sensitive. Second, intermittent supply is hard to measure; service may vary by hour, day, season, and neighborhood, and official "coverage" statistics often mask rationing schedules, low pressure, or system downtime (Kumpel & Nelson, 2016; WHO & UNICEF, 2021). Third, informal provision is inherently difficult to quantify. Vendors, tanker deliveries, kiosks, and shared taps may operate outside formal reporting channels, with prices and quality changing dynamically during shortages (Komives et al., 2005; Allen et al., 2006). Fourth, there are real incentive and governance constraints: documenting unequal service can create political risk for utilities and governments, while disclosing infrastructure layouts can raise security concerns or expose illicit connections (Batley & McLoughlin, 2015).

For geospatial analysis, these constraints mean that urban water mapping often relies on assembling a "best available" mosaic from multiple imperfect sources. At the global or multi-country scale, commonly used building blocks include: (1) household survey and monitoring frameworks such as the WHO/UNICEF Joint Monitoring Programme (JMP), which provide authoritative definitions and trend data but are typically too coarse for neighborhood-scale city mapping (WHO & UNICEF, 2021); (2) facility and point datasets such as the Water Point Data Exchange (WPdx), which compile geolocated water points and attributes where data exist, but have uneven coverage and are most useful as inputs for specific regions rather than a universal city solution (WPdx, 2024); (3) population distribution grids such as WorldPop, which support exposure and service-gap analyses but represent modeled population surfaces rather than direct measures of water access (WorldPop, 2023); and (4) remotely sensed and environmental layers (rainfall anomalies, surface water dynamics, land cover, and elevation) that help characterize hazard context and system stress but cannot directly observe household reliability, cost, or quality (IPCC, 2022).

Because no single global body routinely collects city-scale, neighborhood-level water access data with standardized measures of reliability, affordability, and quality, the most defensible high-resolution maps typically require primary data collection or local administrative integration—for example, household surveys, water-point audits, or utility operational logs (hours of service, pressure zones, break records) linked to a GIS (WHO & UNICEF, 2021; UN-Water, 2023). This is why studies like the Blantyre "walking for water" work are so valuable: they demonstrate what it takes to translate lived experience into spatially explicit, analyzable variables—often at significant cost and logistical effort.

In practice, the implication for this chapter's mapping workflow is straightforward. City-scale water access mapping is rarely a matter of downloading a single "water access layer." It requires (1) defining access variables explicitly (distance, reliability, cost, quality, governance), (2) selecting proxy datasets where direct data are unavailable, and (3) being transparent about uncertainty and bias—especially where informal systems dominate or utility data are incomplete (Komives et al., 2005; Kumpel & Nelson, 2016). The goal is not perfect measurement, but a credible, reproducible spatial model that makes inequities visible and supports better decisions about investment, accountability, and service expansion (Table 4.1).

TABLE 4.1

Dimensions of Water Accessibility and Sources

Variable/ Dimension	Typical Data Sources	Spatial Resolution	Strengths	Limitations
Drinking water access (basic/ improved)	WHO–UNICEF Joint Monitoring Programme (JMP)	National; sometimes subnational	Authoritative definitions; long time series; global coverage	Too coarse for city or neighborhood analysis; does not capture reliability, cost, or intermittency (WHO & UNICEF 2021)
Water point locations	Water Point Data Exchange (WPdx), NGO surveys	Point-level (where available)	Geolocated infrastructure; attributes on functionality and management	Uneven geographic coverage; strongest in sub-Saharan Africa; limited urban completeness (WPdx, 2024)
Utility service areas/ networks	Municipal utilities; regulators	Variable (often polygon or line)	Directly linked to formal infrastructure and governance	Often incomplete, outdated, proprietary, or politically sensitive; rarely indicate service quality (Batley & McLoughlin, 2015)
Reliability/ intermittency	Utility logs; household surveys; academic studies	Neighborhood to household	Captures lived experience of access	Rarely standardized; labor-intensive to collect; often unavailable publicly (Kumpel & Nelson, 2016)
Water quality	National labs; DHS/MICS add-ons; academic sampling	Point-level	Direct health relevance; chemical and microbial indicators	Sparse sampling; expensive; difficult to interpolate reliably (Fewtrell et al., 2005; WHO, 2017)
Cost/ affordability	Household surveys; vendor studies	Household or neighborhood	Reveals regressive pricing and inequality	Highly variable; informal markets difficult to track (Whittington et al., 1991; Komives et al., 2005)
Population distribution	WorldPop; GPW	Grid (100 m–1 km)	Enables exposure and service-gap analysis	Modeled estimates, not census counts; does not indicate access (WorldPop, 2023)
Environmental context	Remote sensing (rainfall, surface water, land cover, DEMs)	Grid (30 m–5 km)	Consistent global coverage; supports hazard analysis	Cannot directly observe access, cost, or reliability (IPCC, 2022)

4.3 Modeling Threat Risk in Geospatial Terms

Mapping water scarcity is not the same as mapping water access. Hydrological models can estimate surface runoff, groundwater recharge, and basin-scale water gaps with increasing sophistication, yet these measures alone do not explain who actually receives water in cities, when it flows, or at what cost. Urban water access is mediated by infrastructure networks, governance arrangements, informal systems, and social power—factors that introduce sharp spatial discontinuities into what might otherwise appear as continuous hydrological processes (Graham & Marvin, 2001; Kumpel & Nelson, 2016).

From a geospatial perspective, this distinction matters. A well mapped as a point feature may indicate the presence of infrastructure, but it does not capture variability in yield, water quality, seasonal reliability, or access restrictions. Similarly, proximity to a pipe or treatment plant does not guarantee service if pressure is low, rationing schedules are in place, or connections are unaffordable. Modeling urban water risk therefore requires shifting the analytical focus from where water exists to how water access is produced and constrained across space (UN-Water, 2023).

This section introduces a set of conceptual and analytical tools for modeling urban water access using GIS. Rather than seeking a single definitive metric, it emphasizes multi-variable approaches that integrate physical supply, infrastructure performance, service reliability, cost, and governance. The goal is not to replicate utility engineering models, but to build spatial representations that make inequities visible, support comparative analysis, and inform decisions about intervention and accountability (Komives et al., 2005; WHO & UNICEF, 2021).

4.3.1 Conceptual Tools for Modeling Urban Water Risk

At a minimum, geospatial models of water access must answer four questions:

1. Where are the water sources and delivery points located?
2. Who can realistically reach and use them?
3. How reliable, affordable, and safe is the service?
4. How do these conditions vary across space and time?

Answering these questions requires combining network analysis, spatial interpolation, service-area modeling, and multi-criteria evaluation—techniques discussed in detail in the sections that follow (Longley et al., 2015; Malczewski, 2006).

4.3.2 Project Design Questions

Before building a geospatial model of urban water access, it is essential to clarify what question the map is meant to answer. Water access is not a single condition, but a combination of distance, reliability, affordability, quality, and governance, each of which may matter differently depending on the analytical goal (WHO & UNICEF, 2021; Komives et al., 2005). A model designed to support emergency response, for example, will prioritize proximity and reliability, while one intended to inform long-term investment may focus on service gaps, infrastructure age, and population exposure.

Scale is equally important. Citywide averages obscure neighborhood-level disparities, while household-scale models may require data that are costly or unavailable. Analysts must also decide whose access counts—formal connections only, or the full spectrum of informal and hybrid systems that supply many urban residents (Allen et al., 2006). Finally, assumptions and uncertainties should be made explicit, particularly where proxy data are used. Thoughtful project design ensures that geospatial models illuminate inequities rather than inadvertently masking them.

4.3.3 Core Variables and Indicators for Modeling Water Access

Geospatial models of urban water access are most effective when they are built around a small number of clearly defined, interpretable variables. While data availability varies widely across cities, five core dimensions recur across the literature and practice of urban water analysis: distance, reliability, cost, quality, and governance (WHO & UNICEF, 2021; UN-Water, 2023). Together, these indicators capture both the physical and institutional conditions that shape access.

Distance is often the most straightforward variable to model spatially. It represents the travel time or walking distance from households to the nearest functional water source, whether a household connection, community tap, kiosk, or vendor. Distance is strongly associated with water consumption levels and time burdens, particularly in cities with intermittent supply or incomplete networks (Thompson et al., 2001; Graham et al., 2016). Network-based distance measures—using road or path networks—are generally more accurate than simple Euclidean buffers.

Reliability refers to how consistently water is available over time: hours per day, days per week, or seasons per year. Intermittent supply is a defining feature of many urban water systems, and it is closely linked to contamination risk and household coping strategies (Kumpel & Nelson, 2016). Although reliability data are rarely available as continuous spatial layers, they can be approximated using rationing schedules, household surveys, or proxy indicators such as elevation and distance from trunk lines.

Cost captures the price paid per unit of water, including both monetary and non-monetary costs. Households reliant on vendors or kiosks often pay significantly more per liter than those with formal connections, making cost a key driver of inequality (Whittington et al., 1991; Komives et al., 2005). Mapping cost surfaces can reveal regressive pricing patterns that are invisible in utility rate schedules alone.

Quality reflects chemical and microbiological safety. While direct water-quality measurements are spatially sparse, point sampling combined with interpolation can identify areas of elevated contamination risk, particularly under intermittent supply conditions (Fewtrell et al., 2005; WHO, 2017).

Finally, governance describes who controls water provision—public utilities, private firms, NGOs, community groups, or informal vendors. Governance shapes pricing, accountability, and investment decisions, making it an essential, if difficult-to-map, dimension of access (Batley & McLoughlin, 2015; Allen et al., 2006).

4.3.4 Lab: Building a Water Access Map with Open Data

This lab demonstrates how open-source data can be combined to create a first-order map of urban water access. The objective is not to produce a definitive assessment, but to model spatial inequalities transparently using widely available inputs.

A typical workflow begins by assembling water point or service data, such as geolocated taps, wells, kiosks, or utility service areas, drawn from sources like WPdx, municipal records, or NGO surveys (WPdx, 2024). These points are paired with population distribution data—for example, WorldPop grids—to estimate how many people fall within specified service distances (WorldPop, 2023). Note that most national statistical organizations around the world collect demographic data at a finer scale and may share it, depending on their policies.

Next, the distance or travel-time analysis is conducted using road or path networks to calculate realistic access zones around water sources. Where reliability data are available, sources can be weighted by hours of service or frequency of outages. In the absence of direct data, analysts may use proxies such as elevation, distance from main lines, or reported rationing schedules (Kumpel & Nelson, 2016).

Additional layers can be introduced to enrich the model. Cost information from household surveys or vendor studies can be joined at neighborhood scales, while water quality sampling points can be interpolated to identify areas of higher health risk (Fewtrell et al., 2005). These variables may be combined using multi-criteria evaluation, producing a composite water access index that reflects multiple dimensions of access rather than proximity alone (Malczewski, 2006).

Throughout the process, assumptions and data limitations should be documented clearly. Open-data water maps are best understood as decision-support tools, not precise measurements, and their value lies in revealing relative disparities and priority areas for further investigation or intervention (UN-Water, 2023).

4.3.5 Advanced Modeling Approaches: Dealing with Uncertainty in Access, and the Utility of Scenario Generation

More advanced geospatial modeling can extend basic access maps to explore system dynamics, uncertainty, and future risk. One important direction involves incorporating network pressure and hydraulics conceptually into spatial analysis. While full hydraulic modeling typically falls outside GIS workflows, pressure zones, elevation gradients, and distance from trunk mains can be used as spatial proxies for service reliability and failure risk (Kumpel & Nelson, 2016; Gleick, 2018).

Another critical challenge is uncertainty. Urban water data are often incomplete, outdated, or biased toward formal systems. Advanced analyses therefore benefit from sensitivity testing—examining how results change when assumptions about reliability, population distribution, or informal provision are varied (Longley et al., 2015). Explicitly mapping uncertainty can be as important as mapping access itself, particularly for policy applications.

Finally, scenario analysis allows analysts to explore how urban water access might change under different futures. Scenarios may include population growth, climate-induced drought, infrastructure investment, tariff reform, or expanded informal provision. By adjusting model inputs to reflect these conditions, GIS can support "what-if" analysis that helps policymakers anticipate trade-offs and unintended consequences (IPCC, 2022; UN-Water, 2023).

Together, these advanced approaches reinforce a central message of this chapter: urban water access is not static. It is produced through interacting physical, institutional, and social processes that can be modeled spatially—but only when uncertainty is acknowledged and equity is treated as a core analytical concern rather than an afterthought.

4.4 Getting to Action on the Water Scarcity Threat

Understanding where and why urban water access fails is only useful if it informs action. Water scarcity is not an abstract risk but a lived condition that demands intervention by utilities, governments, civil society, and residents themselves. This section outlines practical pathways for enhancing resilience in urban water systems, emphasizing how geospatial tools can guide preparedness, response, and long-term reform. While no single strategy applies universally, successful interventions share common principles: transparency, equity, adaptability, and accountability (UN-Water, 2023; WHO & UNICEF, 2021).

4.4.1 Principles for Enhancing Urban Water Resilience

A first principle is anticipation rather than reaction. Cities that wait for reservoirs to reach crisis levels before acting face far higher social and economic costs than those that monitor trends and intervene early. GIS supports anticipation by integrating hydrological data, infrastructure condition, and population exposure to identify neighborhoods most vulnerable to shortages before service collapses (Gleick, 2018; IPCC, 2022).

Second, resilience depends on diversification of supply and demand management. Cities overly dependent on a single reservoir, aquifer, or transfer scheme are inherently fragile. Conservation measures—such as leakage reduction, metering, tiered pricing, and public awareness campaigns—often yield faster and cheaper gains than new supply infrastructure, as demonstrated in Cape Town's response to its Day Zero crisis (Pascale et al., 2020; Brookings Institution, 2019).

Third, equity must be treated as a design criterion, not an afterthought. Interventions that stabilize overall supply can still worsen inequality if they privilege already well-served areas. Targeting investments using spatial access maps helps ensure that network expansion, pressure management, and emergency supply prioritize the most underserved neighborhoods (Komives et al. 2005; Peloso & Morinville 2014).

4.4.2 Planning, Infrastructure, and Technology

At the planning level, GIS supports mapping evacuation-style water response zones, identifying areas that require emergency water delivery during outages or droughts. Utilities increasingly use spatial dashboards to track main breaks, pressure drops, and boil-water advisories in near real time, improving coordination across agencies (OECD, 2016).

Infrastructure interventions remain essential. These include replacing aging mains, expanding storage, rehabilitating well fields, and reducing non-revenue water through leak detection and pressure management. Satellite imagery, mobile data collection, and crowd-sourced reporting now complement traditional asset inventories, allowing utilities to prioritize repairs more strategically (UN-Water, 2023; Kumpel & Nelson, 2016).

In many cities, informal systems must be integrated rather than eliminated. Registering vendors, regulating prices, and monitoring quality can improve safety and affordability while longer-term network expansion proceeds (Allen et al., 2006).

4.4.3 Human Impacts and Adaptive Responses: The Case of Packaged Water

One highly visible adaptive response to unreliable water access is the rise of packaged water markets, such as sachets in West Africa. While often criticized for plastic waste and privatization, research shows that sachet water can improve short-term access to safe drinking water, particularly for children, in contexts where municipal supply is intermittent or unsafe (Stoler et al., 2012; Stoler et al., 2015). This illustrates a broader point: coping strategies are neither purely failures nor solutions. GIS can help evaluate where such adaptations reduce risk—and where they entrench inequity.

4.4.4 Citizen Monitoring, and Forecasting "Day Zero"

Historically, water monitoring has been the domain of state agencies and utilities, but citizen participation is increasingly important. Community reporting of outages, water-quality concerns, and illegal connections can supplement official data, particularly where institutional capacity is limited (UN-Water, 2023). Simple tools—SMS text alerts, WhatsApp groups, or open mapping platforms—can disseminate information rapidly, provided they are trusted and well-managed.

Forecasting future crises remains challenging. Climate models introduce uncertainty, which, while never good, should not be ignored either. Even considering this, studies consistently show that climate change increases the likelihood of extreme drought events similar to Cape Town's Day Zero (Pascale et al., 2020). Rather than predicting exact dates, scenario-based modeling allows cities to test preparedness under plausible futures, strengthening adaptive capacity without over-reliance on precise forecasts (IPCC, 2022).

4.4.5 Looking Ahead: From Scarcity to Governance Reform

Urban water scarcity is not inevitable. It reflects choices about infrastructure investment, pricing, land use, and political priorities. While current trajectories suggest continued strain and uneven access, these outcomes are not fixed. Cities that combine transparent data, geospatial analysis, inclusive governance, and sustained public engagement have demonstrated the ability to avert crisis and reduce inequality (Gleick, 2018; UN-Water, 2023).

Water access is ultimately a governance challenge made visible across space. GIS cannot solve scarcity on its own, but it can expose inequities, justify intervention, and empower communities to demand accountability. In doing so, it transforms water from an invisible (ignored) background service into a public issue that can be debated, mapped, and changed.

The same hydrological processes that shape water access also govern flood risk. Where scarcity threatens cities through absence, floods threaten through excess. The next chapter examines how surplus water becomes a hazard—and how similar tools of mapping, planning, and political action can mean the difference between resilience and catastrophe.

References

Allen, A., Dávila, J. D., & Hofmann, P. (2006). The peri-urban water poor: Citizens or consumers? *Environment and Urbanization, 18*(2), 333–351. https://doi.org/10.1177/0956247806069628

Batley, R., & McLoughlin, C. (2015). The politics of public services: A service characteristics approach. *World Development, 74*, 275–285. https://doi.org/10.1016/j.worlddev.2015.05.018

Brookings Institution. (2019). *Lessons from Cape Town's water crisis*. https://www.brookings.edu/

Davis, J. (2004). Corruption in public service delivery: Experience from South Asia's water and sanitation sector. *World Development, 32*(1), 53–71. https://doi.org/10.1016/j.worlddev.2003.07.003

Fewtrell, L., Kaufmann, R. B., Kay, D., Enanoria, W., Haller, L., & Colford, J. M. (2005). Water, sanitation, and hygiene interventions to reduce diarrhoea in less developed countries: A systematic review and meta-analysis. *The Lancet Infectious Diseases, 5*(1), 42–52. https://doi.org/10.1016/S1473-3099(04)01253-8

Foster, S., & Briceño-Garmendia, C. (2010). *Africa's infrastructure: A time for transformation*. World Bank.

Foster, S., & Chilton, J. (2003). Groundwater: The processes and global significance of aquifer degradation. *Philosophical Transactions of the Royal Society B, 358*(1440), 1957–1972. https://doi.org/10.1098/rstb.2003.1380

Ghana Water Company Limited (2018). *Annual Report 2017–2018*. Accra: GWCL.

Gleick, P. H. (2018). *The world's water: The biennial report on freshwater resources* (Vol. 9). Island Press.

Gough, K. V., & Yankson, P. W. K. (2012). A neglected aspect of the housing market: The care and maintenance of housing in Ghana. *Urban Studies, 49*(3), 561–577. https://doi.org/10.1177/0042098011400772

Graham, S., & Desai, R. (2022). Uneven infrastructures: Urban water, power, and the politics of networked services. In S. Graham & C. McFarlane (Eds.), Infrastructural lives: Urban infrastructure in context. Routledge.

Graham, S., & Marvin, S. (2001). *Splintering urbanism: Networked infrastructures, technological mobilities and the urban condition*. Routledge.

Graham, J. P., Hirai, M., & Kim, S.-S. (2016). An analysis of water collection labor among women and children in 24 sub-Saharan African countries. *PLoS ONE, 11*(6), e0155981. https://doi.org/10.1371/journal.pone.0155981

Grafton, R. Q., et al. (2018). The economics of water scarcity and allocation. *Annual Review of Resource Economics, 10*, 1–20. https://doi.org/10.1146/annurev.resource.050708.144256

Grant, R. (2009). *Globalizing city: The urban and economic transformation of Accra, Ghana*. Syracuse University Press.

Green Dispatch. (2024). *Water scarcity: A growing problem*. https://greendispatch.substack.com/p/water-scarcity-a-growing-problem

IPCC. (2022). *Climate Change 2022: Impacts, adaptation and vulnerability*. Cambridge University Press.

Kingdom, B., Liemberger, R., & Marin, P. (2006). *The challenge of reducing non-revenue water in developing countries*. World Bank.

Komives, K., Foster, V., Halpern, J., & Wodon, Q. (2005). *Water, electricity, and the poor: Who benefits from utility subsidies?* World Bank.

Konikow, L., and Kendy, E. (2005). Groundwater depletion: a global problem. *Hydrogeology Journal, 13*, 317–320. https://doi.org/10.1007/s10040-004-0411-8

Kumpel, E., & Nelson, K. L. (2016). Intermittent water supply: Prevalence, practice, and microbial water quality. *Environmental Science & Technology, 50*(2), 542–553. https://doi.org/10.1021/acs.est.5b03973

Leijnse, M., Bierkens, M. F. P., Gommans, K. H. M., Lin, D., Tait, A., & Wanders, N. (2024). Key drivers and pressures of global water scarcity hotspots. *Environmental Research Letters, 19*(5), 054035. https://doi.org/10.1088/1748-9326/ad3c54

Longley, P. A., Goodchild, M. F., Maguire, D. J., & Rhind, D. W. (2015). *Geographic information science and systems* (4th ed.). Wiley.

Malczewski, J. (2006). GIS-based multicriteria decision analysis: A survey of the literature. *International Journal of Geographical Information Science, 20*(7), 703–726. https://doi.org/10.1080/13658810600661508

McFarlane, C. (2008). Governing the contaminated city: Infrastructure and sanitation in colonial and post-colonial Bombay. *International Journal of Urban and Regional Research, 32*(2), 415–435. https://doi.org/10.1111/j.1468-2427.2008.00793.x

Nilsson, D., & Nyanchaga, E. N. (2008). Pipes and politics: A century of change and continuity in Kenyan urban water supply. *Journal of Modern African Studies, 46*(1), 133–158.

OECD. (2016). *Water governance in cities.* OECD Publishing.

Ofiesh, C. (2011). *Walking for water in Malawi: Modeling travel time to public water points in a GIS-based case study if intra-urban water access.* Master's thesis, Department of Geography, GWU.

Pascale, S., Kapnick, S. B., Delworth, T. L., & Cooke, W. F. (2020). Increasing risk of another Cape Town "Day Zero" drought in the 21st century. *Proceedings of the National Academy of Sciences, 117*(47), 29495–29503. https://doi.org/10.1073/pnas.2009144117

Peloso, M., & Morinville, C. (2014). "Chasing for water": Everyday practices of water access in peri-urban Ashaiman, Ghana. *Water Alternatives, 7*(1), 121–139.

Shiklomanov, I. A., & Rodda, J. C. (2003). *World water resources at the beginning of the twenty-first century.* Cambridge University Press.

Songsore, J. (2008). Environmental health and sanitation in Ghana: A case study of Accra. *Journal of Environment & Development, 17*(4), 426–448.

Sorenson, S. B., Morssink, C., & Campos, P. A. (2011). Safe access to safe water in low income countries: Water fetching in current times. *Social Science & Medicine, 72*(9), 1522–1526. https://doi.org/10.1016/j.socscimed.2011.03.010

Stoler, J., Tutu, R. A., Ahmed, H., & Weeks, J. R. (2012). The relationship between household water insecurity and diarrhoeal disease in Ghana. *International Journal of Environmental Health Research, 22*(5), 456–472.

Stoler, J., Weeks, J. R., & Fink, G. (2015). Sachet drinking water in Ghana's Accra-Tema metropolitan area: Past, present, and future. *Journal of Water, Sanitation and Hygiene for Development, 5*(4), 553–564. https://doi.org/10.2166/washdev.2015.100

Thompson, J., Porras, I., Tumwine, J. K., Mujwahuzi, M. R., Katui-Katua, M., Johnstone, N., & Wood, L. (2001). *Drawers of water II: 30 years of change in domestic water use and environmental health in East Africa.* International Institute for Environment and Development.

UN-Water. (2023). *UN World Water Development Report 2023: Partnerships and cooperation for water.* UNESCO.

UNESCO. (2023). *The United Nations World Water Development Report.*

Wada, Y., Wisser, D., & Bierkens, M. F. P. (2014). Global modeling of withdrawal, allocation and consumptive use of surface water and groundwater resources. *Earth System Dynamics, 5,* 15–40. https://doi.org/10.5194/esd-5-15-2014

Water for People. (2021). *Urban water access and service delivery in Blantyre, Malawi.* https://www.waterforpeople.org/

Water Point Data Exchange (WPdx). (2024). *WPdx data portal.* https://www.waterpointdata.org/

Whittington, D., Lauria, D. T., Mu, X., & Haughton, D. (1991). Paying for urban services: A study of water vending and willingness to pay in Onitsha, Nigeria. *World Development, 19*(2–3), 179–198. https://doi.org/10.1016/0305-750X(91)90254-F

Whittington, D., et al. (2009). Water pricing and affordability: Principles and evidence from low-income countries. *World Bank Research Observer, 24*(2), 275–296.

WHO. (2017). *Guidelines for drinking-water quality* (4th ed.). World Health Organization.

WHO & UNICEF. (2021). *Progress on household drinking water, sanitation and hygiene 2000–2020: Five years into the SDGs*. World Health Organization & UNICEF.

WorldPop. (2023). *WorldPop global population data*. https://www.worldpop.org/

5

Mapping the Danger of Flooding in the City

Chapter Introduction

Flooding is among the most frequent and destructive environmental hazards affecting cities worldwide. Unlike many other hazards, floods often combine sudden physical force with longer-term social and economic disruption. Lives may be lost in minutes, and recovery can take years. In cities in the Global South, flooding is especially damaging because it intersects with informal settlement, inadequate infrastructure, and limited institutional capacity (United Nations Office for Disaster Risk Reduction [UNDRR], 2025; IPCC, 2023).

Hurricane María's impact on Puerto Rico in 2017 illustrates how flooding cascades through urban systems. Beyond immediate inundation, prolonged power outages disrupted water treatment, healthcare delivery, and communications. Floodwaters contaminated drinking supplies, damaged roads and bridges, and isolated communities for weeks. These impacts were unevenly distributed, with low-income and rural populations facing the greatest risks (Kishore et al., 2018; Santos-Burgoa et al., 2018).

Flooding is therefore not simply a hydrologic event. It is a compound hazard whose consequences reflect pre-existing inequalities, infrastructure conditions, and governance choices. Mapping flood risk requires attention not only to where water flows but also to where people live, how cities are built, and who bears the costs of failure.

5.1 Setting the Scene: Urban Flooding

Hurricane María's impact on Puerto Rico in 2017 is considered the deadliest and costliest hurricane to strike the island and archipelago of Puerto Rico in history. Using excess mortality analysis, researchers at George Washington University's Milken Institute School of Public Health estimated that 2,975 deaths were attributable to Hurricane María in Puerto Rico (Santos-Burgoa et al., 2018).

Hurricane María was unlike anything Puerto Rico had seen in generations. As Abigail Matos-Pagán described it, the storm didn't just pass over

DOI: 10.1201/9781032636948-5

the island—it touched every single community, stripping away familiar systems of support and exposing deep structural vulnerabilities. María caused catastrophic damage to homes, health facilities, and basic infrastructure, but it was the complete collapse of communication and power systems that transformed the emergency into a humanitarian crisis. With cell towers down and roads blocked, people could not call for help, and emergency teams often had no way to reach those who needed them most.

Medical care was among the first systems to fail. Hospitals were shuttered, pharmacies were closed, and residents—especially the elderly—were left without access to oxygen, dialysis, insulin, or urgently needed medication. Many of Puerto Rico's older adults live alone, and during María, thousands had no neighbors or family nearby to bring water, food, or medicine. Weeks passed before electricity reached some neighborhoods, and in certain mountainous areas, the power was out for months. The resulting indirect deaths—triggered by lack of medical care, heat exposure, and the collapse of essential services—numbered in the thousands.

Subsequent storms and flooding have only reinforced how precarious daily life can be on the island. In some regions, rainfall exceeded 30 inches, turning streets into rivers and washing away bridges that had only recently been repaired. Flooding repeatedly hits the same communities, often those with the fewest resources to rebuild. Matos-Pagán emphasized that people's cultural ties to land and home often make evacuation impossible or unthinkable; many stay because they have nowhere to go, no guaranteed shelter, and no trust that government services will support them after they leave.

She noted that these events make clear the urgent need to strengthen infrastructure, communication networks, the electric grid, and disaster-response coordination. Puerto Rico needs reliable ways to identify and assist vulnerable residents before a storm strikes—registries, local response teams, neighborhood-based networks of care. And she stressed the importance of recognizing the psychological toll: trauma and chronic uncertainty continue long after the floodwaters recede.

María's devastation is ultimately a story about people, not just hazards. Behind every flooded street and broken bridge were individuals navigating fear, illness, loss, and the struggle to survive without reliable institutions. The hurricane revealed how fragile essential systems can be—and how quickly that fragility becomes lethal when communication, healthcare, and power fail all at once. For cities in the Global South and beyond, the lesson is sobering: resilience is not measured by the height of a levee, but by the strength of the networks—social, infrastructural, and governmental—that protect people when nature pushes past the limits of ordinary life.

This section draws on findings synthesized by the National Academies of Sciences, Engineering, and Medicine regarding community flood impacts, vulnerabilities, and preparedness challenges in Puerto Rico (NASEM, 2024).

5.2 Unpacking the Threat: Defining Flood Danger

Floods are, at their simplest, water where it shouldn't be, since they're defined as the overflow of water (or debris-laden flow) that submerges land normally kept dry. While the dictionary definition is straightforward, legal and technical definitions vary across jurisdictions. Under U.S. federal law (Federal Emergency Management Agency, 2011; Federal Emergency Management Agency, 2019), a flood is officially recognized when two or more acres of normally dry land, or two or more properties, are partially or fully inundated. Other definitions emphasize impact thresholds, such as whether water has damaged property, destroyed crops, or killed domestic animals.

The United Nations Office for Disaster Risk Reduction (UNDRR) reports an annual average loss for flooding of US$388 billion globally (for all floods and infrastructure damage). A 2024 study reports for floods: 4,713 floods were recorded in 168 countries (1990–2022), affecting more than 3.2 billion people, causing 218,353 deaths, and responsible for over US$1.3 trillion in damages. According to David Pogue (2021), floods kill 8,000 people worldwide every year.

Globally, flood policy is vast and context-specific, but most frameworks converge on a single aim: minimizing harm by strengthening society's ability to anticipate, withstand, and recover from inundation. Governments play a central role in this process—through zoning, infrastructure investment, emergency management, and social protection systems that reduce vulnerability. Among the most effective tools are early warning systems, which provide timely, trusted information to communities, including those living in isolated or informal settlements who are often the least able to protect themselves when water rises.

In mountain regions, elevation and slope are among the strongest predictors of flood risk and damage, a reality reflected in Colorado's well-known "Climb to Safety" campaign. The state's Fluvial Hazard Zone (FHZ) mapping program is accessible through the web-based Colorado Hazards and Risk Map Portal, and it emphasizes that traditional FEMA floodplain maps, which assume a static river channel and model only water inundation, fail to capture the full spectrum of dangers (Colorado Geological Survey, 2022).

Using LiDAR-derived terrain models and detailed geomorphic mapping, the Colorado Water Conservation Board (2023) and the Colorado Fluvial Hazard Mapping Program show that slope, channel confinement, and valley geometry are essential for identifying where floodwaters and debris flows will accumulate or where they may undercut roads, bridges, or other infrastructure (Colorado Water Conservation Board, 2023). Taken together, Colorado's approach demonstrates why mountainous cities, towns, and villages—whether in the Rockies, Andes, or Himalayas—must account for geomorphic processes alongside standard inundation models when assessing risk.

At its core, flooding is a function of physics and climate. Water, like all matter, follows gravity—moving downhill through surface channels and groundwater systems until it pools in low-lying areas and eventually returns to the ocean. In the atmosphere, water vapor forms part of the global hydrologic cycle, continuously evaporating, condensing, and returning as precipitation. Under stable conditions, floods are part of this background rhythm of exchange; but as climate change injects more heat and energy into the oceans and atmosphere, the water cycle accelerates, driving more intense storms and heavier downpours. The result is not a new process, but an amplified one—the same flows of surface and groundwater pushed beyond historical limits. A simple hydrologic typology distinguishes floods by source—fluvial (riverine), pluvial (rainfall-driven), and coastal (storm surge or sea-level rise)—but all share the same root mechanism: too much water arriving faster than it can be absorbed or conveyed away.

Figure 5.1 illustrates tidal "nuisance" flooding, which is often amplified by sea-level rise. In the photo, a man pushes a car in a flooded section of road after a big rain event.

FIGURE 5.1
Nuisance flood on street. (By Sveta K, on Pexels. Creative Commons CC—Free to Use. https://www.pexels.com/photo/flood-on-street-8568719/)

TABLE 5.1

Impacts of Floodwaters in Cities

Impact Type	Mechanism/Process	Typical Indicators or Outcomes	Example City/Region
Scouring and Erosion	Fast, turbulent flow undermines foundations and transport structures	Exposed pilings, collapsed culverts, and road subsidence	Accra (Bridge abutment failures, 2022)
Velocity and Hydraulic Force	High-energy water ($\geq 2\,m/s$) carries vehicles, debris, and people	Vehicles displaced, building collapse in low-lying streets	Freetown, Sierra Leone—2017 flash flood-landslide event
Sediment and Bedload Transport	Mobilized silt and debris clog drainage and raise riverbeds	Clogged culverts and reduced channel capacity	Manila, Philippines
Contaminant Spread	Floodwaters mix with sewage, industrial effluents, and solid waste	Outbreaks of cholera, leptospirosis, and skin infections	Dhaka, Bangladesh
Thermal and Biochemical Effects	Warm standing water accelerates bacterial growth and decay	Elevated post-flood disease incidence, odor, and fish kills	Lagos, Nigeria
Infrastructure Short-Circuiting	Water intrusion damages electrical, transport, and telecom systems	Grid failures, loss of mobility, and communication blackouts	Mumbai, India
Subsurface Instability	Prolonged saturation weakens soils, causes slope or foundation failure	Landslides, sinkholes, cracked walls and pavements	Manizales, Colombia
Salinization and Intrusion	Coastal surges push saltwater into aquifers and farmland	Soil infertility, saline groundwater, and damaged crops	Maputo, Mozambique
Displacement and Housing Loss	Informal housing on marginal land destroyed by inundation	Emergency shelters, internal migration, and property loss	Port-au-Prince, Haiti
Public Health Burden	Standing water creates mosquito breeding and increases waterborne disease risk	Rise in malaria, dengue, and cholera after floods	Phnom Penh, Cambodia
Economic Disruption	Market closures, supply-chain breakdown, and informal labor halted	Drop in daily income, long recovery period	Karachi, Pakistan
Education and Gendered Impacts	Flood recovery diverts women and children from school or work	Reduced attendance, increased caregiving burden	Jakarta, Indonesia

Source: Compiled by the author from case studies and reports cited in this chapter, including United Nations Office for Disaster Risk Reduction (2022), UN-Habitat (2020), Douglas et al. (2008), and related disaster reporting and urban flood studies.

5.2.1 How Do Floods Affect Cities?

Urban expansion can often worsen flooding when there's a lack of stormwater management infrastructure, i.e., no storm sewer system. And, the physical processes of flood management are overlaid with human systems, prior histories of injustice or neglect, land reclamation, informal housing on floodplains, blocked drains, and poor maintenance of infrastructure. Climate change compounds all of these by intensifying rainfall and raising sea levels.

In practice, urban resilience is less a one-time achievement than a continuous act of maintenance. It depends on everyday labor—people who clear storm drains, repair pumps, and keep floodgates and culverts functional before and after the rain. True resilience requires a steady corps of municipal workers who are adequately trained and ideally unionized and protected with fair wages, pensions, and life insurance. But it also depends on capital—public investment in drainage upgrades, maintenance budgets, and transparent contracting processes that prevent neglect from becoming catastrophe. In this sense, resilience is inseparable from good governance: cities that sustain strong institutions of care, accountability, and public works are the ones best able to absorb and recover from floods.

At first glance, effective flood management might seem to demand massive engineering works such as concrete channels, retention basins, spillways, and outwash plains on the model of Los Angeles or Houston in the United States. These "gray" systems are built to convey water away quickly, but they are enormously expensive to construct and maintain. They often transfer risk downstream rather than reducing it overall. In many cities, resilience may depend less on megaprojects than on smarter, distributed systems: well-maintained drains, restored wetlands, and responsive governance that keeps infrastructure functioning through routine care. The lesson from Los Angeles is not simply to build bigger channels—it's to understand the hydrology of the city as a living system that must be managed continuously, not conquered once.

Cairo, the capital of Egypt and a metropolis of more than 20 million people, has made substantial investments in water management infrastructure since the 1950s, during a political era that promoted rapid urbanization through large-scale public works—especially in water supply, irrigation, and flood management. The Nile River has always been the axis of these investments: its annual flow supports agriculture, drinking water, and industry across the Nile Valley and Delta. The Nile Delta, where roughly 60 percent of Egypt's food is produced and where Cairo and Alexandria together form a continuous urban corridor, remains both the country's economic heartland and one of its most climate-vulnerable regions (Hereher, 2015).

5.2.2 Storm Related Flooding with Growing Impacts: Convective Storms

The example of West Point, New York, illustrates how convective storms intensify beyond the capacity of existing systems. As that flood event pointed out, no engineered soft surface can withstand that volume of water. On July 9, 2023, torrential thunderstorms dumped an estimated 8 inches (about 200 mm) of rain in a matter of hours over the West Point and Highland Falls area of the Hudson Valley, triggering an official Flash Flood Emergency. The volume overwhelmed hillside runoff controls and campus drainage infrastructure at the U.S. Military Academy, leading to landslides, basement flooding, washed-out roads and a multibillion-dollar repair effort by the U.S. Army Corps of Engineers.

On May 10–11, 2022, parts of Guangzhou and Shaoguan in the People's Republic of China recorded 275 mm (10.8 inches) of rainfall in just 3 hours, the highest 3-hour total since records began. The China Meteorological Administration classified it as a super-cell convective event triggered by strong instability and high precipitable water content rather than monsoon inflow. Urban underpasses and metro tunnels flooded within minutes, overwhelming drainage infrastructure.

Advanced technologies, such as weather stations, flow meters, satellite and LiDAR sensing, and drone-based damage assessments, can offer extraordinary tools for measuring, modeling, and predicting floods. These systems, however, remain concentrated in the Global North and are often treated as enhancements rather than essentials, while many cities still lack basic hydrometric data.

Even the most sophisticated technology cannot guarantee safety. In October 2024, the region around Valencia, Spain, experienced a "DANA" (*Depresión Aislada en Niveles Altos*) (aka cold-drop) event that dumped extreme rainfall, a record 184.6 mm (7.25 inches) in 1 hour in Turís, in the province of Valencia. Although the national weather agency issued some red alerts, authorities misjudged the storm's intensity, some warnings were late or downplayed, and storm intensity was underestimated, resulting in catastrophic losses in several high-risk municipalities. The ensuing flood killed more than 228 people, revealing that information systems alone do not make a city resilient, and that preparedness and governance matter just as much as technology. Disasters happen in places that aren't prepared for them, and they can exacerbate tensions in informal communities.

5.2.3 Discussion: The Threat of Floods versus the Threat of Formalization

Disasters like floods do not strike people and regions evenly. They expose and amplify pre-existing vulnerabilities, particularly in informal communities where risk is already normalized. Flooding in such settings is rarely just a physical hazard; it becomes a catalyst that intensifies social tension, governance failure, and difficult trade-offs between safety and displacement.

The experience of the Vietnam barrio (deprived area) in San Juan illustrates this dilemma clearly. Reporting on the neighborhood in the aftermath of Hurricane María emphasized not only its exposure to flooding and landslides but also its long history of resilience that preceded the disaster. For decades, residents built and rebuilt homes in the absence of formal recognition, navigating mudslides, bureaucratic uncertainty, and chronic neglect. What emerged was a durable community shaped by necessity and mutual support. Post-disaster recovery, however, reframed this endurance as a liability. Even as large sums were allocated for reconstruction, residents faced pressure to relocate in the name of safety, revealing a fundamental tension between risk reduction and the right to remain (New York Times (2019) reporting; see also Santos-Burgoa et al., 2018).

From a geographic perspective, the Vietnam barrio occupies a position of compounded disadvantage. It is an improvised and unregulated settlement situated within an island territory that itself occupies an ambiguous political status relative to the United States. Scale matters here: neighborhood-level vulnerability is nested within broader colonial, fiscal, and infrastructural constraints. Throughout this book, informal settlements recur as sites where failures of governance—pirated electricity lines, clogged drainage channels, and unstable housing—are highly visible, while less visible but equally important social assets such as flexible networks, shared labor, and local knowledge often go unrecognized.

Why, then, do some cities recover quickly from floods while others experience prolonged decline? Flooding rarely occurs in isolation. It clusters with other chronic urban stresses such as inadequate housing, air pollution, unreliable infrastructure, and weak public institutions. These conditions reinforce one another, creating what might be described as a syndrome of urban vulnerability rather than a single hazard problem.

A comparison with Acapulco, Mexico, highlights how recovery priorities shape post-flood trajectories. After Hurricanes Otis (2023) and John (2024), recovery efforts focused heavily on restoring coastal tourism infrastructure as a driver of economic revival. International visibility and revenue generation were treated as prerequisites for broader recovery, even as many residents continued to face disrupted services and housing insecurity. This strategy reflects a different political calculus: rebuilding what sustains external investment first, with uneven implications for local vulnerability (World Bank, 2023).

Together, these cases underscore a central tension in flood risk management. For informal communities, the threat of flooding is often inseparable from the threat of formalization—where safety interventions themselves become mechanisms of displacement. Effective resilience, therefore, cannot be reduced to engineering solutions or relocation mandates alone. It requires rebuilding trust, addressing structural inequality, and recognizing that risk reduction strategies themselves redistribute power, security, and belonging.

At every scale, floods produce winners and losers, even within the same neighborhood. Some of this differentiation reflects straightforward topography: wealthier households are often located on higher ground, while poorer or excluded populations occupy lower, more flood-prone areas. These patterns are not new. In industrial-era London, health inequalities were shaped less by elevation than by prevailing wind patterns. Westerly winds carried coal smoke eastward, leading wealthier households to cluster in the cleaner western districts, while working-class populations were concentrated in the more polluted east. This spatial sorting produced one of the earliest documented geo-epidemiological gradients, with higher exposure to coal smoke associated with increased infant mortality and reduced life expectancy (e.g., Hanlon, 2020).

In cities in the Global South today, similar gradients emerge through what might be described as the urbanization of poverty. Rapid urban growth, limited formal housing supply, and weak enforcement of land-use regulations push low-income households into marginal land—often floodplains, wetlands, or poorly drained peripheries. A case study from Bobo-Dioulasso, Burkina Faso, for example, documents how low-lying informal settlements experience recurrent flooding, blocked drainage, and heightened disease risk during the rainy season, despite strong social cohesion and adaptive coping strategies (Douglas et al., 2008; UN-Habitat, 2020).

These examples underscore a central point: flood risk is rarely distributed randomly. It reflects long-standing spatial inequalities that operate at multiple scales and across historical periods. In the next section, we turn to how these patterns can be modeled explicitly using geospatial tools.

5.3 Modeling Flood Risk in Geospatial Terms

Flood modeling brings together topography, hydrology, and exposure data. Popular YouTuber and GIS educator Hans van der Kwast demonstrates how to model and visualize flood risk in Rotterdam using entirely open-source tools and datasets. Working in QGIS, he overlays OpenStreetMap (OSM) building footprints and road networks with national elevation data from the AHN (Actueel Hoogtebestand Nederland)—a high-resolution LiDAR-based digital elevation model (DEM). By applying the Raster Calculator and "less than" threshold filters to simulate inundation depths (for example, water rising 1–3 m above mean sea level), he generates flood-extent polygons that reveal which neighborhoods, streets, and critical facilities would be underwater under different sea-level scenarios.

The workflow illustrates the power of open, replicable modeling. With free global basemaps and national elevation layers, students or planners can replicate the analysis for any deltaic or low-lying city. Van der Kwast emphasizes not only the technical steps (DEM clipping, reclassification, symbolization) but also the communication aspect—how cartographic clarity and interactivity can turn raw data into persuasive, policy-relevant visualizations.

At minimum, a workable flood-risk GIS project draws on a small set of foundational spatial layers that describe terrain, water movement, and the built environment. These typically include a DEM to capture slope and flow direction; hydrographic data representing rivers, canals, and drainage networks; land-use or land-cover information, particularly measures of impervious surface; and some representation of floodplains or historical flood extents. Together, these layers provide a first-order picture of where water is likely to accumulate, how it will move across the urban landscape, and which people and assets may be exposed when systems are overwhelmed.

5.3.1 More Advanced Flood Modeling

Kumar et al. (2023), in their flood modeling review article, show that the way models work is through simplification: real-world complexity is reduced to mathematical inputs (such as precipitation, temperature, topography and slope, soil type, and land cover) that can then be run to estimate flood impacts.

In a nutshell, hydrologic models use mathematical tools and computer simulations to understand and predict water's behavior in systems like rivers, watersheds, and groundwater. Geographic inputs such as those mentioned above simulate the flow rate, volume, and timing of water at different sites. Hydrologic modeling produces data that are used in hydraulic modeling, which simulates how water would behave during a flood event. Hydraulic modeling examines how water will flow through a system in the surrounding terrain, accounting for channel shape, roughness, and water velocity. The depth and extent of flooding in each area, as well as the location and timing of peak flows, are better assessed with hydraulic modeling, which can be coupled with hydrologic modeling to leverage the benefits of both approaches. Frequently used software packages used for hydraulic modeling include HEC-RAS, MIKE FLOOD, and TUFLOW.

The next model type covered by Kumar et al. is the numerical flood model. Numerical flood models are computer-based models that use math and computational techniques to simulate the behavior of water during a flood event. They use numerical algorithms to solve equations taking into consideration rainfall, runoff, channel geometry, and riverbed roughness. Large-scale simulations demonstrate the benefit of numerical models over physical flood models, if you want to do a large-scale simulation.

Rainfall runoff models are hydrological models that simulate the transformation of rainfall into runoff in a catchment. This approach is important for predicting the amount and timing of runoff in a catchment for early warning or evacuation purposes, and it can be empirical, conceptual, or physical.

TABLE 5.2

Conceptual Tools and Analytics for Urban Flood Mapping

Tool/Approach	What It Does	Common Uses, Strengths, and Limits
Flood Risk Mapping	Integrates terrain, hydrography, and historical flood information to delineate areas likely to be inundated	Useful for baseline risk screening and public communication. Results are sensitive to DEM resolution and often simplify flow dynamics, especially in dense urban areas
Flow Accumulation Analysis	Models how surface water converges across terrain based on slope and drainage paths	Effective for identifying runoff pathways and potential problem areas such as low points and blocked drains. Assumes surface flow dominance and may underrepresent subsurface or engineered drainage systems
Scenario Modeling	Simulates hypothetical flooding conditions such as extreme rainfall, sea-level rise, or infrastructure failure	Supports planning under uncertainty and comparison across futures. Scenarios are illustrative rather than predictive and depend heavily on assumptions about water levels and system behavior
Critical Infrastructure Overlay	Intersects flood extents with the locations of roads, hospitals, utilities, schools, and other essential services	Highlights cascading risk and service disruption. Data availability and accuracy often vary widely across cities, particularly in informal areas
Early Warning and Monitoring Systems	Visualizes sensor networks, rainfall gauges, river stages, and alerts in near real time	Enables preparedness and rapid response where monitoring exists. Effectiveness depends on institutional capacity, maintenance, and public trust in warnings

Source: Compiled by the author based on the flood-modeling and geospatial analysis literature cited in this chapter, including Kumar et al. (2023) and related GIS-based flood risk studies.

The next category of models are ones based on remote sensing and GIS. These can be used to analyze and assess topography, hydrology, meteorology, and land use, and they can include satellite imagery and remote sensing.

The integration of machine learning and AI, autonomously identifying patterns and trends in the data plus real-time monitoring, can simulate patterns for identifying objects in high-resolution imagery and so on. Kumar et al. (2023) review how machine learning and AI methods are now being used to autonomously identify flood-prone patterns in high-resolution imagery—such as classifying inundated areas, detecting drainage blockages, and mapping debris flows—from low-cost and open-source data. The authors note that these techniques enable "ungauged" or poorly monitored basins, which are often found in low-income cities, to benefit from near-real-time flood-monitoring and hazard-mapping capabilities that were previously available only to well-instrumented watersheds.

5.3.2 Project Design Questions

Designing a flood-mapping project requires translating observations of the local environment into geographic space in ways that are both analytically sound and practically feasible. That translation depends not only on the hazard itself but also on data availability, technical capacity, and the purpose of the analysis. Before selecting tools or running models, it is useful to clarify the basic structure of the workflow: what information will be assembled first, how intermediate products will be generated, and how results will ultimately be evaluated and communicated.

A critical early step is to inventory the core datasets available for the study area. Most flood analyses rely, at minimum, on a DEM, hydrographic data, land-use or land-cover information, and some representation of impervious surfaces. Building footprints, precipitation records, and population data may further refine exposure estimates, but their availability and reliability vary widely across cities. The question is not whether a dataset exists somewhere, but whether it is appropriate in resolution, date, and geographic extent for the scale of the problem being examined.

Project design also involves trade-offs between speed, complexity, and certainty. In many real-world settings, a "quick and dirty" analysis using readily available data can be more valuable than a technically sophisticated model that depends on hard-to-obtain inputs or opaque assumptions. Analysts must decide which variables matter most for the question at hand and which uncertainties can be tolerated, particularly when working under time pressure or data scarcity.

Finally, flood maps are never neutral outputs. Every project tells a story about risk, responsibility, and consequence. Whether the goal is early warning, infrastructure planning, community advocacy, or policy debate, the usefulness of the map depends on how clearly it conveys what is at stake and for whom. Effective project design therefore requires not only technical judgment but also narrative judgment—an understanding of how spatial analysis can inform decisions rather than merely describe hazards.

5.3.2.1 Lab 5.1: Building a Simple Flood Inundation Map with Open Data

5.3.2.1.1 Purpose of the Lab

This lab introduces a "quick and dirty" approach to flood inundation modeling using open data and open-source GIS tools. The goal is not to produce a hydraulically precise flood forecast, but to illustrate how elevation data can be used to explore relative exposure under different flooding scenarios. The lab uses San Juan, Puerto Rico, as a running example and models potential inundation associated with incremental sea-level rise.

This approach is commonly used in early-stage risk screening, public communication, and comparative analysis—particularly in contexts where detailed hydrologic models or engineering data are unavailable.

5.3.2.1.2 Conceptual Approach

This lab models four hypothetical flooding scenarios corresponding to 1, 2, 3, and 5 m of sea-level rise. For each scenario, areas at or below the specified elevation are identified as potentially inundated. The results illustrate how exposure expands with increasing water levels and how elevation structure shapes risk. At the most calamitous flooding, 5 m, much of the downtown is covered with water, as shown in Figure 5.2.

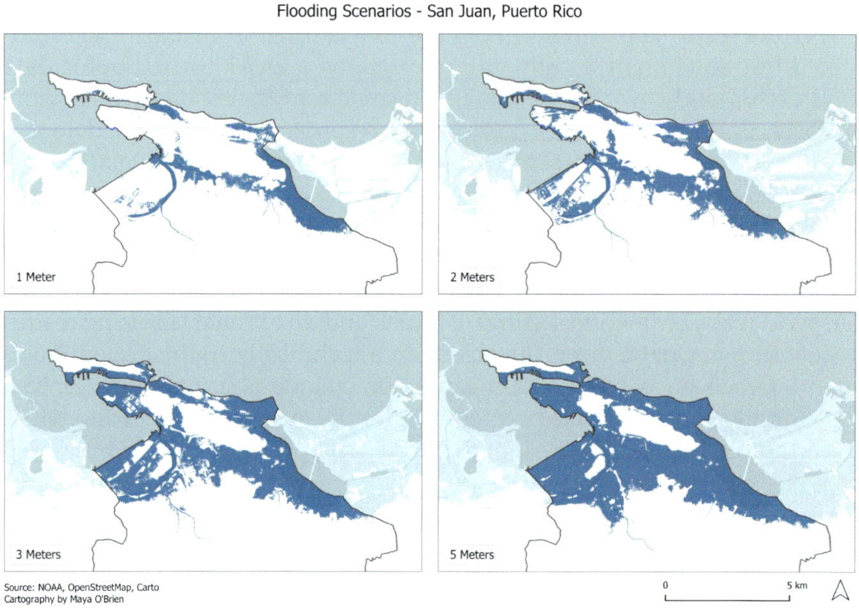

FIGURE 5.2
Example of flood inundation maps for 1, 2, 3, and 5 m of sea-level rise in San Juan, Puerto Rico. (Cartography by Maya O'Brien.)

TABLE 5.3
Core Requirements and Data for LAB 5.1

Category	Component	Purpose and Notes
Software	QGIS (v. 3.30 or later)	Open-source GIS platform used for raster processing, visualization, and basic spatial analysis. Later versions improve raster handling and interface stability
Hardware	Standard laptop or desktop computer	Sufficient for moderate-resolution raster datasets. Very high-resolution DEMs may require additional memory
Connectivity	Internet access	Required to download elevation data, boundaries, and optional population layers

(Continued)

TABLE 5.3 (*Continued*)

Core Requirements and Data for LAB 5.1

Category	Component	Purpose and Notes
Core Spatial Data	Digital elevation model (DEM)	Forms the analytical foundation of the lab by representing terrain, slope, and relative elevation. Resolution and acquisition date strongly influence results
	Study area boundary	Used to clip the DEM and constrain the analysis to a meaningful geographic extent
Optional Exposure Data	Population data	Enables simple exposure estimates by intersecting flood scenarios with residential populations
Common DEM Sources	NOAA (U.S.), Copernicus DEM, NASADEM	Publicly available elevation datasets suitable for screening-level flood analysis; verify resolution and vertical accuracy
Boundary Sources	UN OCHA HDX, national open-data portals, OpenStreetMap	Provide administrative or planning boundaries appropriate for local-scale analysis
Population Sources	U.S. Census, WorldPop, National Statistical Agencies	Choose a spatial unit and year consistent with the scale and purpose of the analysis

Source: Compiled by the author based on standard open-source GIS workflows and publicly available data sources referenced in this chapter.

5.3.2.2 Workflow Overview

5.3.2.2.1 Step 1: Prepare the Data

1. Load the DEM, study area boundary and any optional population layers into QGIS.
2. Reproject all layers into a common coordinate reference system.
3. Clip the DEM to the extent of the study area boundary.

5.3.2.2.2 Step 2: Create the Flood Scenario (1 m)

1. Open the Raster Calculator.
2. Create a conditional expression identifying all DEM cells with elevation $\leq$ 1 m.
3. The output raster will contain binary values:
   - 1 = potentially flooded
   - 0 = not flooded

5.3.2.2.3 Step 3: Convert to Vector Format

1. Polygonize the binary raster.
2. Delete polygons with a value of 0.
3. The remaining polygons represent areas potentially flooded under the 1-m scenario.

5.3.2.2.4 Step 4: Symbolize and Contextualize

1. Symbolize the flood polygons (e.g., blue fill at ~80% opacity).
2. Add a basemap beneath the flood layer to provide geographic context.

5.3.2.2.5 Step 5: Repeat for Additional Scenarios

1. Repeat Steps 2–4 for 2-, 3-, and 5-m flooding scenarios.
2. Create a separate map for each scenario.

5.3.2.3 Optional Extension: Estimating Population Exposure

This optional step introduces a simple exposure analysis.

1. Load population data for the study area.
2. Filter population features to the study area boundary.
3. Use Extract by Location to identify population units intersecting each flood scenario.
4. Calculate:
   - total population in the study area
   - population intersecting each flood polygon
5. Compare exposure across scenarios using a summary table.

5.3.2.4 Interpreting Results

This method produces *indicative*, not predictive, results. It assumes static water levels and does not account for drainage, barriers, or flow dynamics. Its strength lies in comparative insight: identifying which neighborhoods become exposed as water levels rise and how quickly risk expands across the urban landscape.

5.3.3 Economic and Institutional Dimensions

Good flood data are only the start. Your results can be used to make decisions and take action. Looking ahead 5, 10, or even 20 years, we will all see big changes in our local and regional climates with multiple manifestations of climate impact, including hotter summers, shorter and milder winters, and arid areas receiving less rainfall, with the trend being that the dry gets drier and the wet will be wetter. Humid tropical areas will get more heat and more pronounced seasonal drought in monsoon areas, and there will be significant sea level rise.

Flood control should be considered an investment in future safety. Cities need maintenance budgets, building codes, and local enforcement. Many flood-mitigation projects fail not for technical but institutional reasons, including the lack of follow-up, data silos, or politicized decision-making. A cost–benefit analysis can show that building proactive green infrastructures (retention ponds, mangrove buffers) is often cheaper than reactive rebuilding. Investments in multi-use facilities, such as community centers that can double as cooling centers or disaster safe zones, will pay off in multiple ways.

But they have to be paid for. Floating public bonds are a method of financing enhancements to protect facilities and investments from flood losses. It's considered public sector but is enabled through investors buying bonds with a guaranteed rate of return; it is slow but secure. It can finance the construction of parks that can absorb some of the flood waters threatening nearby homes.

If we compare the cost of drainage infrastructure versus post-disaster reconstruction, a World Bank cost–benefit review finds that investing in resilient drainage and flood-mitigation infrastructure now typically costs only a fraction of what post-flood reconstruction and economic losses would incur later—often yielding benefit–cost ratios of two or higher (Hallegatte et al., 2019).

In general, for understanding the impacts of flooding, it's important to highlight the importance of institutional strength, particularly when it affects public investment in infrastructure. For the strength of institutions, how many national governments have dedicated offices concerned with emergency management?

163 countries (about 84% of UN Member States) reported via the Sendai Framework Monitor in 2024 that they have systems reporting on disaster risk reduction indicators. The UNDRR in its Global Assessment Report 2025 notes that although 131 countries reported having a national Disaster Risk Reduction strategy in place by 2023, the report also flags that many of these agencies struggle with execution capacity, data gaps, institutional coordination, and sustainable funding.

A useful term for the ability of neighborhoods and individual housing units in a city to age well over time and be strong in the face of environmental and technological threats is durability. Structure durability is defined here as the long-term ability of built assets (e.g., drainage, flood walls, bridge supports) to withstand repeated hazard loads and return to function without full reconstruction. Prashar et al. (2023) and Nwogu (2025) refer to the broader concept of "infrastructural resilience" and "structural measures" in flood risk GIS modeling.

The social and environmental contexts for floods vary greatly. Durable urban environments can be close-knit, internally coherent, and well-connected to the outside world. Sometimes, the most durable neighborhoods in a city are the older parts that were built first and on the best quality land, a more strategic location, with views and defensibility. These older and more durable quarters tend to be more resilient.

Old San Juan, Puerto Rico, illustrates how infrastructural resilience can emerge not from modern engineering, but from historical siting and durable construction practices. The sixteenth-century walled city sits on a limestone promontory roughly 10–15 m (~33–49 feet) above sea level, buffered by thick masonry fortifications and narrow cobbled streets designed to shed rain quickly toward the harbor. Despite exposure to repeated hurricanes—most recently María (2017) and Fiona (2022)—its elevation, masonry building stock, and compact urban fabric have allowed it to withstand flooding and storm surge far better than newer low-lying suburbs built on reclaimed coastal wetlands. The contrast underscores a key insight of resilience planning: structural durability and advantageous topography can provide passive protection that modern infrastructure often fails to match.

Old urban cores (central business districts) are often extremely well-built, and informal or deprived areas surrounding the formal areas tend to be the opposite of durable and subject to all sorts of environmental threats. Residents in these areas are often stymied by lack of options. How institutions react versus respond more long term (adapt) to hazardous conditions. The level of public investment is often related more to recovery than resilience. "Resilience Capital" approximates social capital, natural capital, and infrastructural resilience. Community knowledge about disasters and flexibility in responding contribute to resilience. Extended families with ties to the countryside can help city dwellers cope.

In the Global South, more hazardous conditions tend to be found in populated areas with industry interspersed, also with a high level of short-term tenancy, water and air pollution, and numerous code and zoning violations. Despite the handicaps, places like this can knit together to withstand outside threats. Often the residents of these areas have high operational environmental knowledge but lack formal representation in government. Citizens and affected people can mobilize through conventional democratic means, through their member of parliament, alderman, ward head, etc., or through advocacy organizations like Slumdwellers International. Their exclusion will have impacts on durability and resilience.

5.4 Getting to Action on Floods

Flood risk reduction requires moving deliberately from data to knowledge and from knowledge to action. The following principles highlight practical pathways for translating flood information into durable urban resilience strategies.

PRINCIPLES FOR ENHANCING URBAN FLOOD RESILIENCE

MAKE RISK VISIBLE AND INTELLIGIBLE

Flood resilience begins with transparency. Mapping exposure and communicating risk through open dashboards, participatory mapping, and citizen science helps residents understand not only where flooding occurs but also why it affects some neighborhoods more than others. When communities can see and validate risk information, trust in both data and institutions tends to improve.

PLAN FOR UNCERTAINTY RATHER THAN A SINGLE FUTURE

Cities can no longer rely on one design standard or forecast. Scenario-based planning allows decision-makers to test adaptation strategies across a range of plausible futures, from incremental sea-level rise to extreme rainfall events. Flexible "adaptation pathways" help keep options open as conditions change and new information emerges.

COMBINE ENGINEERED SYSTEMS WITH NATURE-BASED SOLUTIONS

Green infrastructure—such as bioswales, permeable pavements, wetlands, and floodable parks—can absorb, delay, and redistribute runoff while delivering co-benefits for air quality, biodiversity, and urban livability. The most effective flood strategies typically blend gray and green systems rather than relying exclusively on either.

INSTITUTIONALIZE PREPAREDNESS THROUGH PRACTICE AND DOCUMENTATION

Preparedness cannot be improvised during a storm. Regular drills, scenario exercises, and systematic after-action reporting strengthen institutional memory and reduce response times. Cities that treat emergency planning as a routine civic function, rather than an exceptional activity, are better able to learn from past floods and adjust accordingly.

BUILD DURABLE PARTNERSHIPS ACROSS SECTORS AND SCALES

Flood resilience is co-produced. Collaboration among local governments, universities, civil society organizations, and affected communities helps translate data into action and sustain political momentum. Networks that connect cities to one another further accelerate learning and allow successful approaches to travel across contexts.

5.4.1 Case Illustration: From Dashboards to Action in Freetown

Freetown's Transform Freetown initiative illustrates how open data and strong municipal leadership can translate information into action. Launched in 2019 by the Freetown City Council, the plan identifies priority sectors across resilience, human development, healthy cities, and mobility, and emphasizes performance dashboards and community-driven investment in urban services (Freetown City Council, 2019).

As part of this agenda, the city has used satellite imagery, open-data systems, and tax-base reform to mobilize resources for flood mitigation measures, including drain clearance, hillside stabilization, and nature-based interventions such as the "#FreetownTheTreeTown" campaign. These efforts demonstrate how data-enabled governance can move beyond visualization toward implementation (Lincoln Institute of Land Policy, 2022).

5.4.2 Other Technology Use: The Diffusion of Stream Monitoring

The United States' model of real-time, continuous stream gauging that is led by the U.S. Geological Survey remains a global benchmark for flood monitoring. However, replicating this model in many other parts of the world is constrained by limited institutional capacity, incomplete telemetry, and gaps in long-term data continuity. In countries such as Ethiopia and India, many stream gauges either lack real-time discharge reporting or are unevenly distributed across basins. A regional assessment of the global river gauge network found significant spatial bias in Asia, with monitoring stations clustered in accessible or economically significant locations rather than distributed according to hydrologic need. As a result, many flood-prone basins remain under-monitored, limiting early warning and risk modeling capacity (Krabbenhoft et al., 2022). Figure 5.3 shows a water gauge station on a small rise next to a flowing stream.

FIGURE 5.3
Stream Gauge Photo. 1,600×1,200 pixels. https://upload.wikimedia.org/wikipedia/commons/4/40/B62doddieburn.jpg?download.

These constraints raise a broader planning question: when resources are limited, should cities prioritize technological expansion or retain flexible financial reserves for emergency response and recovery?

5.4.3 Green Infrastructure and Nature-Based Solutions

Flood resilience can be enhanced through green infrastructure such as wetlands, retention ponds, reservoirs, and bioswales that provide space for excess water. Recent studies identify bioswales and vegetated swales as increasingly recognized nature-based solutions for urban stormwater and flood-risk management, though adoption varies widely depending on institutional support and maintenance capacity (Fletcher et al., 2024).

The implementation challenges remain significant. Design guides emphasize that many bioswale projects underperform due to unclear maintenance responsibility, insufficient funding, or inadequate integration with drainage systems (NACTO, 2017). Recent World Bank assessments note that while ecosystem-based adaptation is receiving increased support, detailed project-level evidence on bioswales remains limited, and many investments continue to favor conventional gray infrastructure such as levees and floodwalls (World Bank, 2025).

Remote sensing research further highlights the role of ecosystems in flood protection. Studies using satellite and LiDAR data show how mangrove restoration contributes to coastal resilience by enhancing wave attenuation, carbon storage, and storm surge protection, linking ecosystem health directly to flood-risk reduction (Fatoyinbo, 2025).

5.4.4 Does Your City Have a Flood Impact Mitigation Plan?

Although thousands of cities worldwide have experienced damaging flood events, empirical research suggests that only a minority have articulated comprehensive flood mitigation or adaptation plans. A global study of 9,468 cities across 175 countries found highly uneven adoption of structured flood-risk management actions, with significant gaps in low- and middle-income regions (Gandhi et al., 2022). At present, no comprehensive global registry of cities with formal flood mitigation plans exists, complicating efforts to benchmark preparedness or track progress systematically.

5.4.5 Building Durable Flood-Mitigation Infrastructure

In cities in the Global South, building durable flood-mitigation infrastructure increasingly requires moving beyond conventional gray systems toward hybrid approaches that combine engineered drainage with natural retention and absorption. Such systems include floodable parks, retention basins, wetlands, and restored waterways that provide space for water while delivering social and ecological co-benefits.

Financing these approaches remains a central challenge. Effective implementation depends on access to new financial instruments, project preparation capacity, and institutional partnerships that can sustain investments over time. Comparative research highlights the importance of sequencing visible, community-scale interventions—such as the creation of multi-use inundation areas or neighborhood-scale green infrastructure—to build trust and establish a track record that enables broader adaptation goals (Iliadis et al., 2023). Complementary measures, including tenant protections, community organizations, and a free press, play an important role in supporting neighborhood stability and ensuring that adaptation does not become a pathway to displacement.

Yet even the most carefully designed and well-financed flood-mitigation systems have limits, and in some locations, repeated exposure, rising seas, or compounding hazards eventually force a more difficult question: not how to protect a place, but whether it can safely remain inhabited at all.

5.4.6 When Staying is No Longer Possible: Relocation of Communities

While often politically and emotionally difficult, the relocation of entire communities has been implemented successfully under certain conditions. A growing body of research emphasizes that outcomes depend less on engineering precision than on social process, institutional capacity, and legitimacy. Comparative analyses of disaster-induced relocation highlight several recurring principles that shape whether relocation reduces long-term risk or reproduces vulnerability (Balachandran et al., 2025).

5.4.6.1 Ground Decisions in Transparent Risk Science

Relocation should be anchored in hazard assessments that communities can understand, scrutinize, and question. When risk determinations are opaque or framed as purely technical judgments, they tend to erode trust and invite resistance. Transparent risk science—clearly communicated and open to contestation—helps distinguish unavoidable exposure from politically convenient displacement.

5.4.6.2 Prioritize Consent and Participation

Relocation efforts are more durable when affected communities participate meaningfully in decision-making. Co-designing relocation plans—by selecting sites, layouts, phasing, and timelines—strengthens legitimacy and improves long-term stability. Participation is not simply a procedural requirement; it is a practical mechanism for aligning technical solutions with lived realities.

5.4.6.3 Protect Place Attachment and Social Continuity

Displacement severs more than physical ties to land. Cultural practices, social networks, and collective memory are often rooted in specific places. Relocation strategies that preserve proximity, naming, spatial layouts, or communal institutions can reduce trauma and help sustain identity, particularly where communities have occupied marginal land for generations.

5.4.6.4 Treat Relocation as a Long-Term Process, Not an Event

The move itself marks only the beginning of adaptation. Successful relocations require sustained institutional support well beyond construction, including access to livelihoods, services, governance structures, and conflict resolution mechanisms. Without long-term engagement, relocated communities may face renewed vulnerability or secondary displacement.

5.4.6.5 Confront Historical and Political Context Directly

Relocation unfolds within existing power relations shaped by land tenure systems, colonial legacies, and prior injustices. These histories often resurface during resettlement, influencing who bears risk and who benefits from safety interventions. Ignoring the political context can undermine even technically sound plans, while acknowledging it can open space for repair and redistribution.

Taken together, these principles suggest that relocation outcomes hinge less on site selection or housing design than on the quality of governance and social process that accompany them. In this sense, relocation is not merely a technical response to flood risk, but a profound reconfiguration of place, authority, and belonging.

5.4.7 Memorializing Lost Places

Relocation also raises enduring questions about memory, identity, and loss. When communities move, places do not simply disappear; they acquire new meanings shaped by absence, remembrance, and contested futures. Decisions about what happens to abandoned sites—whether they are preserved, commemorated, repurposed, or allowed to erode—shape how displacement is understood both locally and historically.

The island community of Gardi Sugdub (Guna Yala, Panama) is widely cited as one of the first climate-driven relocations in Central America. Rising seas and recurrent flooding rendered the island increasingly uninhabitable, prompting the collective relocation of approximately 300 Guna families to a mainland settlement. While the move reduced immediate physical risk, it also introduced profound challenges related to cultural continuity and the status of the island itself.

Gardi Sugdub illustrates that relocation does not end when people move. The fate of abandoned places—whether they remain sites of memory, ritual, and identity or fade from collective attention—continues to shape how communities experience loss and resilience in the context of climate change (AP News reporting; UNDRR, 2025).

5.4.8 Lessons from Dominíca

The Caribbean island of Dominíca, severely affected by Hurricane María in September 2017, offers an instructive example of post-disaster recovery aligned with long-term resilience. The storm caused an estimated US$1.4 billion in losses, but subsequent recovery efforts emphasized integrated development rather than piecemeal reconstruction.

Assessments of Dominíca's recovery highlight several transferable lessons for other small island developing states. These include embedding capacity building within project design to address chronic skill shortages; ensuring rapid access to liquidity following disasters to stabilize households and local economies; and establishing strong project coordination units capable of managing multisector resilience initiatives. Together, these elements allowed Dominíca to couple disaster recovery with broader institutional strengthening, rather than treating reconstruction as a return to the status quo.

5.4.9 Bottom Line

Flooding is not going away. As climate change intensifies hydrologic extremes, cities face growing pressure to adapt under conditions of uncertainty, inequality, and constrained resources. Technical solutions alone are insufficient. Effective flood resilience depends on governance systems capable of learning over time, maintaining infrastructure, earning public trust, and confronting difficult trade-offs about protection, retreat, and loss.

In this sense, flood mapping is not merely an exercise in prediction or visualization. It is a tool for collective decision-making—one that can clarify who is at risk, what is valued, and how societies choose to live with water in an increasingly unstable climate.

References

Balachandran, B., Olshansky, R. B., & Johnson, L. A. (2025). *Planning for disaster-induced relocation of communities: Lessons from 53 cases* (Working paper). University of Illinois Hazard Planning Research Program.

Colorado Geological Survey. (2022). *Case study: Fluvial hazard zone mapping*. Colorado Department of Natural Resources.

Colorado Water Conservation Board. (2023). Colorado fluvial hazard zone (FHZ) mapping program. Colorado Department of Natural Resources. https://coloradohazardmapping.com

Douglas, I., Alam, K., Maghenda, M., McDonnell, Y., McLean, L., & Campbell, J. (2008). Unjust waters: Climate change, flooding and the urban poor in Africa. *Environment and Urbanization*, *20*(1), 187–205.

Fatoyinbo, L. (2025). Preserving Earth's wetlands with satellite technology. *National Geographic Society News*. https://news.nationalgeographic.org/lola-fatoyinbo-preserving-wetlands-with-satellite-technology/

Federal Emergency Management Agency. (2011). *A whole community approach to emergency management: Principles, themes, and pathways for action*. FEMA.

Federal Emergency Management Agency. (2019). *Hurricane María mitigation assessment team report*. FEMA.

Fletcher, T. D., et al. (2024). Nature-based solutions for stormwater and flood-risk management. *Water Research*, *248*, 120833.

Freetown City Council. (2019). Transform Freetown: Transforming lives through inclusive climate resilience and human development (2019–2022). Freetown City Council. https://fcc.gov.sl/wp-content/uploads/2019/01/Transform-Freetown-Agenda.pdf

Gandhi, V., Kahn, M. E., & Walsh, R. (2022). Adapting to flood risk: Evidence from a panel of global cities (NBER Working Paper No. 29930). National Bureau of Economic Research. https://doi.org/10.3386/w29930

Hallegatte, S., Rentschler, J., & Rozenberg, J. (2019). Lifelines: The resilient infrastructure opportunity. World Bank. https://doi.org/10.1596/978-1-4648-1430-3

Hanlon, W. W. (2020). Coal smoke, city growth, and the costs of the Industrial Revolution. *The Economic Journal*, *130*(630), 462–488.

Hereher, M. (2015). Vulnerability of the Nile Delta to sea level rise: An assessment using remote sensing. Geomatica, Natural Hazards and Risk, 6(3), 203–214. https://doi.org/10.1080/19475705.2013.845115

Iliadis, P., Markonis, I., & Tsakiris, G. (2023). A cost-benefit source–receptor framework for implementation of blue-green flood risk management. arXiv. https://doi.org/10.48550/arXiv.2311.00420

Intergovernmental Panel on Climate Change. (2023). *Climate change 2023: Synthesis report*. IPCC.

Kishore, N., Marqués, D., Mahmud, A., Kiang, M. V., Rodriguez, I., Fuller, A., Ebner, P., Sorensen, C., Racy, F., Lemery, J., Maas, L., Leaning, J., Irizarry, R. A., Balsari, S., & Buckee, C. O. (2018). Mortality in Puerto Rico after Hurricane Maria. The New England Journal of Medicine, 379(2), 162–170. https://doi.org/10.1056/NEJMsa1803972

Krabbenhoft, C. A., Allen, G. H., Lin, P., Godsey, S. E., Allen, D. C., Burrows, R. M., DelVecchia, A. G., Fritz, K. M., Shanafield, M., & Zimmer, M. A. (2022). Assessing placement bias of the global river gauge network. Scientific Data, 9, 758. https://doi.org/10.1038/s41597-022-01677-9

Kumar, V., Sharma, K. V., Caloiero, T., Mehta, D. J., & Singh, K. (2023). Comprehensive overview of flood modeling approaches: A review of recent advances. *Hydrology*, *10*(7), 141. https://doi.org/10.3390/hydrology10070141

Lincoln Institute of Land Policy. (2022). *Urban resilience and local finance in Freetown*. Lincoln Institute of Land Policy.

National Academies of Sciences, Engineering, and Medicine. (2024). Exploring community flood impacts, vulnerabilities, and adaptation strategies to public health concerns: Proceedings of a workshop—in brief. Washington, DC: The National Academies Press.

National Association of City Transportation Officials. (2017). *Urban street stormwater guide*. NACTO.

New York Times. (2019, January 20). Kimmelman, M. Rebuilding Puerto Rico after Hurricane Maria. *The New York Times*.

Nwogu, N. (2025). Integrating GIS into flood risk management. *Sustainability*, *17*(3), 1021.

Pogue, D. (2021). *How to prepare for climate change*. Simon & Schuster.

Prashar, N., et al. (2023). Urban flood resilience: A comprehensive review. *Journal of Hydrology*, *620*, 129442.

Rodríguez, M. I., & Santiago-Báez, L. (2020). Cultural heritage and urban resilience in San Juan, Puerto Rico. *Sustainability*, *12*(17), 6789.

Santos-Burgoa, C., et al. (2018). Differential and persistent risk of excess mortality from Hurricane Maria in Puerto Rico. *The New England Journal of Medicine*, *379*(2), 162–170. https://doi.org/10.1056/NEJMsa1803972

UN-Habitat. (2020). *World cities report 2020: The value of sustainable urbanization*. UN-Habitat.

United Nations Office for Disaster Risk Reduction. (2025). *Global assessment report: Flood hazard profile*. UNDRR.

World Bank. (2025). *Strengthening new infrastructure assets: A cost-benefit analysis*. World Bank.

6

Fire and Technological Dangers

Chapter Introduction

If water represents one class of sudden danger for cities, fire and technology represent another; both are hazards that tend to erupt violently and often without warning. Fires can spread through industrial zones or informal settlements in minutes. Technological failures such as gas leaks, chemical releases, and electrical faults can turn ordinary urban spaces into sites of mass injury or long-term contamination. The same forces that drive urban density also concentrate fuels and hazardous materials. When government systems, including infrastructure maintenance and emergency services, fail to keep pace with rapid urban growth, these hazards become more frequent.

Fire and technological dangers also expose the deeply uneven geography of risk (Wisner et al., 2004, UN-Habitat, 2020, WHO, 2016). In many cities around the world, residents rely on informal energy sources in overcrowded housing, dramatically increasing the likelihood of ignition. Industrial facilities handling toxic or flammable materials are often sited close to dense neighborhoods, reflecting weak zoning, regulatory capture, or the prioritization of economic growth over public safety. Climate variability further amplifies these risks by drying vegetation, stressing power systems, and increasing demand for cooling and energy. Together, these dynamics reveal that urban fires and technological disasters are rarely isolated accidents. Instead, they are the spatial expression of historical and contemporary land-use decisions, making them especially well-suited for geospatial analysis and intervention.

6.1 Setting the Scene

In August 2023, one of the deadliest wildfires in U.S. history swept through the historic town of Lahaina on West Maui, Hawaii, transforming a vibrant coastal community into a landscape of ash, rubble, and loss. Fueled by exceptionally dry conditions, strong seasonal winds, and abundant flammable grasses, the blaze raced through residential neighborhoods and down commercial streets with astonishing speed. Official tallies recorded 102 confirmed

DOI: 10.1201/9781032636948-6

deaths, with additional people missing and thousands displaced from their homes. By the time the flames were extinguished, nearly 2,200 buildings had been destroyed, erasing generations of cultural heritage and livelihoods and leaving survivors to confront the shock of profound physical and emotional devastation (State of Hawai'i Office of the Governor, 2023; National Weather Service, 2023).

This disaster did not occur in a vacuum. The environmental and economic backstory of West Maui shaped both the fuel on the landscape and the vulnerability of the community. Until the late nineteenth century, the broader Lahaina region supported a dense indigenous Hawaiian population rooted in a sustainable agricultural system of taro patches, fishponds, and flowing streams that nourished crops and people alike. Over decades of colonial and capitalist transformation, this system was reconfigured to support large-scale sugar plantations and other extractive enterprises. After the closure of the West Maui sugar plantation in 1999, water that once irrigated diversified crops was increasingly diverted or left unmanaged, and active cultivation that had historically kept invasive grasses and shrubs in check largely ceased. The result was an accumulation of highly flammable, non-native vegetation across hillsides that now border urban and peri-urban zones. Contemporary climate variability—manifested in prolonged dry spells and high temperatures—added another layer of stress, drying out fuels and lowering the threshold for ignition.

When drought conditions intensified in the summer of 2023, these legacy land uses and climatic forces converged. In the absence of sufficient moisture, swaths of non-native grasses and brush became tinder, ready to burn. Seasonal wind patterns—enhanced by shifts in atmospheric pressure—then propelled the fire front through landscapes both wild and developed, making suppression efforts extraordinarily difficult. Emergency responders faced the dual challenge of controlling rapidly spreading flames and conducting search-and-rescue operations in neighborhoods that caught fire and burned faster than many residents could flee. For many survivors, the fire's ferocity felt sudden and inexplicable. But for geographers and hazard specialists, the conflation of ecological transformation, water management histories, invasive fuel loads, and climate stressors underscored that such events are not purely "natural disasters," but the unfortunate outcomes of long-term socio-environmental processes.

The Lahaina wildfires expose the interplay between historical land-use change, ecological vulnerability, and rapid onset hazard events. In cities and peri-urban zones worldwide, from informal settlements in the Global South to affluent suburbs of wealthy nations, the legacies of past economic and environmental policies continue to shape present-day risk. Understanding those pathways is essential not just for reconstructing what happened in West Maui in 2023, but for rethinking how communities plan for, respond to, and ultimately reduce the risk of similarly catastrophic fires in the decades ahead.

6.2 Unpacking the Fire and Tech Hazard Threat

Fire and technological hazards occupy a distinctive position among urban dangers because of their capacity to overwhelm response systems almost instantaneously. Unlike slow-onset threats such as water scarcity or chronic air pollution, fires and industrial accidents often emerge from routine urban activities and escalate rapidly once ignition or system failure occurs. In cities, these hazards are rarely random. They reflect spatial arrangements that concentrate flammable materials, hazardous substances, and vulnerable populations within dense built environments (Wisner et al., 2004; UNDRR, 2022).

At a basic level, urban fire risk arises from the interaction of fuel, ignition, and exposure. Fuel includes vegetation at the urban fringe, combustible building materials, stored fossil fuels, plastics, textiles, and industrial chemicals. Ignition sources range from open flames and faulty wiring to industrial processes, machinery sparks, and vehicle accidents. Exposure is shaped by population density, building design, evacuation capacity, and proximity to hazardous facilities. When these elements align, particularly where regulatory systems and emergency services lag behind urban growth, fires and technological accidents become predictable outcomes of urban form and governance rather than exceptional events (Burby, 2006; Cutter et al., 2008).

The sections that follow disaggregate fire and technological hazards by type and mechanism before turning to their cumulative health burdens and implications for spatial modeling.

6.2.1 Types of Fire Hazards

Fire hazards in cities can be broadly categorized into wildland and peri-urban fires, urban residential fires, industrial fires, and transportation-related fires. Each category is shaped by distinct spatial drivers.

Wildland and peri-urban fires occur at the interface between vegetated landscapes and human settlement, commonly described as the wildland–urban interface (WUI). Fire spread in these settings is strongly influenced by topography, vegetation type, weather, and infrastructure layout. Fires move more rapidly uphill; slope and aspect affect fuel moisture; and strong winds accelerate fire fronts while transporting embers over long distances (Radeloff et al., 2005; Syphard et al., 2017). As cities expand into hillsides, forests, and savannas, often without commensurate land-use controls, the WUI becomes a zone of heightened and recurring risk. The 2023 Lahaina fire illustrates how invasive grasses, prolonged drought, and wind can transform peri-urban landscapes into rapid fire corridors with devastating consequences.

Urban residential fires are closely tied to everyday household practices. In many Global South cities, residents rely on kerosene, charcoal, wood, or liquefied petroleum gas for cooking and heating. Overloaded electrical circuits, illegal power connections, candles, and open flames are common ignition sources, particularly in informal settlements (WHO, 2016; Bonjour et al., 2013). While individual incidents are often small, their cumulative impact is substantial. Dense housing, narrow access roads, and limited water pressure frequently impede firefighting efforts, allowing localized fires to spread rapidly (UN-Habitat, 2020).

6.2.2 Fire Hazards in Cities

Fire hazards in cities can be broadly categorized into wildland and peri-urban fires, urban residential fires, industrial fires, and transportation-related fires, each with distinct spatial drivers.

Wildland and peri-urban fires occur at the interface between vegetated landscapes and human settlement, often referred to as the WUI. Fire spread in these settings is strongly influenced by topography, vegetation type, weather, and infrastructure layout. Fires move more rapidly uphill; slope and aspect affect fuel moisture; and strong winds can accelerate fire fronts and transport embers over long distances (Radeloff et al., 2005; Syphard et al., 2017). As cities expand into hillsides, forests, and savannas—often without commensurate land-use controls—the WUI becomes a zone of heightened and recurring risk. The 2023 Lahaina fire illustrates how invasive grasses, prolonged drought, and wind can transform peri-urban landscapes into rapid fire corridors with devastating consequences.

Urban residential fires are closely tied to everyday household practices. In many Global South cities, residents rely on kerosene, charcoal, wood, or liquefied petroleum gas for cooking and heating. Overloaded electrical circuits, pirated power connections, candles, and open flames are common ignition sources, particularly in informal settlements (WHO, 2016; Bonjour et al., 2013). Figure 6.1 depicts a typical informal electrical connection. While individual incidents are often small, their cumulative impact is substantial, contributing to burns, smoke inhalation injuries, and preventable deaths. Dense housing, narrow access roads, and limited water pressure frequently impede firefighting efforts, allowing localized fires to spread rapidly (UN-Habitat, 2020).

In Figure 6.2, children play on a trash pile that is partially caught on fire in a refugee camp in the Democratic Republic of Congo.

Industrial fires present a different but equally severe threat. Chemical plants, textile factories, refineries, warehouses, and informal workshops often store large quantities of flammable or toxic materials. In cities with weak zoning or regulatory enforcement, such facilities are frequently located adjacent to dense residential areas. Fires in these contexts can trigger explosions, release toxic plumes, or contaminate soil and water systems. Repeated garment factory fires in Dhaka, Bangladesh, including the 2012 Tazreen Fashions disaster, highlight how global supply chains externalize safety risks onto workers and surrounding communities (Donaghey & Reinecke, 2018; Alamgir & Banerjee, 2019).

Transportation-related fires involve fuel tankers, rail depots, ports, pipelines, and aviation facilities. Accidents at fuel depots or tanker explosions on

FIGURE 6.1
Informal electrical connections. (Jos1950, CC BY-SA 4.0 <https://creativecommons.org/licenses/by-sa/4.0>, via Wikimedia Commons)

congested urban roads can cause mass casualties within seconds. Because transportation corridors often pass through low-income or industrial districts, these hazards disproportionately affect marginalized populations living near highways, ports, or rail yards (OECD, 2018).

6.2.3 Technological Hazards and Fire

Technological hazards overlap with fire risk but extend beyond combustion alone. Gas leaks, electrical failures, chemical releases, and industrial spills may occur independently or in combination with fires. In many disasters,

FIGURE 6.2
Urban residents burning trash. (Location: DRC, Sept 26, 2014. Oxfam East Africa, CC BY 2.0 <https://creativecommons.org/licenses/by/2.0>, via Wikimedia Commons)

fire acts as the trigger that transforms a technical failure into a large-scale public health emergency. Burning plastics, toxic fire suppressants, and industrial by-products can contaminate air, water, and food systems long after the flames are extinguished, producing health impacts that persist for years (Landrigan et al., 2018).

The Bhopal gas disaster remains a defining example of technological risk in an urban context. Although not a fire, the release of methyl isocyanate (MIC) gas from a pesticide plant in 1984 exposed hundreds of thousands of residents to a lethal toxin, with health impacts persisting decades later. Investigations identified poor maintenance, disabled safety systems, weak oversight, and the siting of hazardous industry near dense populations as central causes. These conditions remain common in many rapidly urbanizing regions (Dhara & Dhara, 2002; Varma & Varma, 2005). The Bhopal disaster is examined in more detail later in this chapter.

6.2.4 Climate Variability as a Risk Multiplier

Climate variability and climate change act primarily as risk multipliers rather than sole causes of urban fires and technological disasters. Rising temperatures dry vegetation and strain electrical infrastructure. Heat waves drive increased use of fans and air conditioners, often on circuits not designed to handle sustained high loads, increasing the likelihood of electrical fires.

Drought reduces water availability for firefighting, while extreme wind events accelerate fire spread and complicate evacuation (IPCC, 2022).

These climatic pressures interact with existing urban vulnerabilities, including aging infrastructure, informal housing, and limited emergency response capacity. Together, they lower the threshold at which routine failures escalate into crises.

Despite its toll, fire is often underrepresented in urban risk discourse. Many incidents are treated as localized accidents rather than systemic failures, and emergency response data are frequently fragmented or underreported, particularly in informal settlements. As a result, fires rarely receive sustained policy attention until a catastrophic event forces recognition (UNDRR, 2022).

Global injury and mortality data suggest that fire-related burns, smoke exposure, and toxic inhalation contribute significantly to urban disease burdens. The Global Burden of Disease project estimates that burns alone account for millions of disability-adjusted life years annually, with impacts concentrated among low-income populations (Yin et al., 2025). Fire risk, like many urban hazards, is therefore shaped less by chance than by where people live, work, and accept exposure.

6.2.5 Emerging Topic: Skyscrapers Catching Fire

High-rise fires represent a growing but underappreciated urban hazard, particularly in rapidly urbanizing cities where vertical construction is outpacing fire safety regulation and enforcement. While skyscrapers are often associated with efficiency and density, fires in tall buildings pose unique challenges related to height, evacuation complexity, façade materials, and reliance on mechanical safety systems that may fail under stress (NFPA, 2019; CTIF, 2021).

One of the defining risks of contemporary high-rise fires is the widespread use of combustible exterior cladding, especially aluminum composite panels with polyethylene cores. These materials, valued for their low cost and ease of installation, have been implicated in several catastrophic fires, most notably the 2017 Grenfell Tower fire in London, which killed 72 people. Investigations showed how fire spread rapidly up the building exterior, bypassing internal fire compartmentalization and overwhelming evacuation routes (Hackitt, 2018; Lane, 2020). Similar cladding-related fires have occurred in Dubai, Melbourne, Shanghai, and cities across the Global South, often with limited public accountability (CTIF, 2021).

Firefighting technology has clear vertical limits. Aerial ladders and water pressure systems are often ineffective beyond a certain height, placing heavy reliance on internal sprinklers, standpipes, and smoke-control systems. In many cities, particularly outside high-income countries, these systems are poorly maintained or inconsistently inspected (UN-Habitat, 2020). Evacuation itself becomes hazardous as stairwells fill with smoke, elevators are disabled, and residents may be trapped far above ground.

Climate and energy pressures further exacerbate these risks. Heat waves increase electrical demand, raising the likelihood of electrical fires in buildings with outdated wiring. At the same time, modern high-rise construction relies heavily on synthetic interior materials that burn hotter and release more toxic smoke than traditional materials, increasing mortality from inhalation rather than burns (WHO, 2016; Landrigan et al., 2018).

From a governance perspective, skyscraper fires expose sharp inequalities in urban safety. Luxury towers often incorporate redundant suppression systems and professional maintenance, while mid-range and informal high-rise housing lacks comparable protections. In many Global South cities, fire safety regulations exist on paper but are weakly enforced due to limited inspection capacity or political pressure to accelerate construction (UNDRR, 2022).

CASE BOX: GARMENT FACTORY FIRES IN DHAKA, BANGLADESH

Dhaka, Bangladesh, has become one of the world's most prominent centers of garment manufacturing, anchoring a global apparel supply chain that employs millions and supplies major brands across Europe and North America. This industrial concentration, however, has been accompanied by a persistent pattern of factory fires and building disasters that expose the human cost of weak regulation, rapid urban industrialization, and globalized production pressures (Donaghey & Reinecke, 2018; Alamgir & Banerjee, 2019).

One of the deadliest incidents occurred in November 2012 at the Tazreen Fashions factory, located in an industrial district on the outskirts of Dhaka. A fire broke out on the ground floor of the multi-story building, reportedly triggered by an electrical short circuit. Locked exits, barred windows, and the absence of functional fire escapes trapped workers on upper floors. At least 112 people were killed, many from smoke inhalation or after jumping from the building to escape the flames (ILO, 2013; Ahmed et al., 2014). Survivors described overcrowded production floors, flammable materials stacked near workstations, and a lack of fire drills or emergency training.

The Tazreen disaster was not an isolated event. Bangladesh has experienced hundreds of garment factory fires over the past several decades, often involving similar conditions: overloaded electrical systems, combustible textiles, narrow stairwells, and minimal enforcement of safety codes (Reinecke & Donaghey, 2015). These risks are intensified by the spatial embedding of factories within dense urban neighborhoods, where zoning laws are weak and industrial and residential uses frequently intermingle. Fires that begin inside factories can quickly threaten nearby housing, shops, and informal settlements.

Global economic pressures play a central role in shaping this risk environment. Bangladesh's garment industry operates on thin profit margins, driven by intense competition among suppliers and price pressures from international brands. In this context, investments in fire suppression systems, building retrofits, or worker training are often treated as discretionary costs rather than essential protections. Factory owners may find it cheaper to pay fines or bribes than to undertake comprehensive safety upgrades, particularly where inspection regimes are under-resourced or corruptible (Alamgir & Alakavuklar, 2020).

In the wake of the Tazreen fire and the subsequent Rana Plaza building collapse in 2013, which killed more than 1,100 workers, international attention forced reforms. The Accord on Fire and Building Safety in Bangladesh and the Alliance for Bangladesh Worker Safety introduced independent inspections, public reporting of hazards, and requirements for remediation. These initiatives led to measurable improvements in factory safety conditions, including electrical upgrades and the installation of fire exits and alarms (ILO, 2018). However, enforcement remains uneven, and many smaller factories operating outside formal supply chains continue to pose serious risks.

From a geospatial perspective, Dhaka's garment fires illustrate how industrial hazard risk is shaped by land-use patterns, building layouts, and global economic linkages. Mapping factory locations alongside population density, road access, and emergency response infrastructure reveals clusters of extreme vulnerability, in areas where a single ignition event can result in mass casualties. These fires are not accidents in any meaningful sense; they are the foreseeable outcome of placing hazardous industrial activities within dense urban environments without adequate spatial planning or accountability.

A key takeaway is that garment factory fires in Dhaka demonstrate how fire risk in Global South cities is produced at the intersection of urban form, regulatory weakness, and global supply chains, making them a critical case for geospatial analysis and intervention.

6.2.6 Global Fire Burden and Disability-Adjusted Life Years

Fires rarely dominate headlines outside of major disasters, yet their cumulative global health burden is substantial and persistent. Estimates from the Global Burden of Disease project show that burn injuries alone account for millions of disability-adjusted life years annually, with the highest burdens concentrated in low- and middle-income countries (Yin et al., 2025). Mortality captures only a fraction of fire's true impact. Survivors often experience long-term disability, including chronic respiratory illness, mobility impairment, disfigurement, and psychological trauma.

Fire-related health burdens are unevenly distributed across space and social groups. Children, women, and low-income urban residents face disproportionate risk, particularly where cooking and heating rely on open flames or solid fuels (WHO, 2016). Informal settlements and overcrowded housing amplify exposure, while limited access to emergency medical care increases the likelihood that injuries become fatal or permanently disabling. In many residential fires, smoke inhalation rather than burns is the leading cause of death, linking fire risk directly to broader patterns of urban air pollution (Landrigan et al., 2018).

DALYs provide a critical lens for understanding fire as a chronic urban threat rather than a series of isolated accidents. While individual fires may be small, their aggregate contribution to lost life years rivals that of better-recognized hazards in many cities. Yet fire prevention and burn care remain underfunded relative to their impact, particularly in the Global South (UNDRR, 2022).

6.2.7 From Understanding Threat to Modeling Risk

Unpacking the causes and consequences of urban fires and technological hazards reveals a central challenge. While these threats are well understood qualitatively, they remain unevenly measured, mapped, and managed across cities. Fire risk follows discernible spatial patterns shaped by land use, infrastructure age, building materials, population density, and proximity to hazardous activities. In many cities, particularly in the Global South, these patterns remain obscured due to fragmented data and limited analytical capacity (UNDRR, 2022; UN-Habitat, 2020).

This gap between knowledge and action underscores the need for geospatial modeling as a core tool of urban fire risk management. Fires and technological accidents are inherently spatial phenomena. They ignite at specific locations, spread through landscapes and building networks, and affect populations unevenly depending on access, mobility, and response time. Geographic information systems provide a framework for integrating these dimensions and shifting analysis from retrospective documentation toward prevention (Cutter et al., 2008).

6.2.8 Case Box: Bhopal—The Technological Disaster That Never Ended

On the night of December 2–3, 1984, a pesticide plant operated by Union Carbide India Limited in Bhopal, India, released an estimated 40 metric tons of MIC gas into the surrounding city. The dense toxic cloud spread rapidly through nearby neighborhoods, exposing hundreds of thousands of residents while they slept. Immediate death estimates ranged from 3,000 to 8,000, with long-term mortality and morbidity pushing total casualties far higher over subsequent decades (Dhara & Dhara, 2002; Varma & Varma, 2005).

Although not a fire, the Bhopal disaster remains one of the most consequential technological hazards in urban history, illustrating how industrial failures can produce impacts comparable to or exceeding those of catastrophic fires. Subsequent investigations revealed that multiple safety systems at the plant had been disabled or poorly maintained, including refrigeration units, gas scrubbers, and flare towers designed to neutralize toxic releases. Plant operators lacked adequate training in hazard awareness and emergency preparedness, and surrounding communities were given little information about risks or response procedures (Varma & Varma, 2005). Figure 6.3 shows an exterior view of the Union Carbide pesticide factory in Bhopal, India, the scene of the world's worst industrial disaster in 1985.

A central spatial factor in the disaster was land-use incompatibility. The Bhopal plant was originally sited on the outskirts of the city, but rapid urban growth brought dense residential neighborhoods directly adjacent to the facility. Weak zoning enforcement and regulatory oversight allowed hazardous industrial operations to coexist with civilian housing, effectively embedding extreme technological risk within the urban fabric (Dhara & Dhara, 2002). When the gas was released, proximity became lethal.

FIGURE 6.3
Bhopal factory photo by Julian Nyča, CC BY-SA 3.0 <https://creativecommons.org/licenses/by-sa/3.0>, via Wikimedia Commons. Date: March 15, 2010. (Source: Self-photographed. Author Julian Nyča.)

The health consequences of Bhopal have proven remarkably persistent. Studies document elevated rates of respiratory disease, neurological impairment, reproductive health problems, and cancers among exposed populations decades after the event. Recent spatial difference-in-differences analyses show that in utero exposure to the gas resulted in long-term reductions in human capital, including lower educational attainment and lifetime earnings. These effects extend the disaster's legacy across generations (Bharadwaj et al., 2014). Environmental contamination of soil and groundwater near the plant site continues to pose risks, exacerbated by incomplete remediation.

Analyses of Bhopal consistently converge on a set of lessons relevant to cities worldwide. Industrialization without concurrent investment in safety regulation dramatically increases disaster risk. International standards, even if imperfectly enforced, can establish critical baselines for hazard management. National and local governments play an indispensable role in accident prevention, land-use planning, and emergency preparedness. Most importantly, hazardous industrial facilities should not be located near dense urban populations without robust safeguards and transparent risk communication.

From a geospatial perspective, Bhopal underscores how technological disasters are shaped by where hazards are placed, who lives nearby, and how risks are governed over time. Like urban fires, the Bhopal gas tragedy was not an unforeseeable accident. It was the outcome of identifiable spatial decisions, regulatory failures, and economic priorities—making it a paradigmatic case for understanding technological danger in cities across the planet.

6.2.9 GIS Applications: Mapping Technological Risk and Long-Term Exposure

A growing body of open data related to Bhopal, including environmental sampling, health outcomes, and demographic indicators, is now available through academic repositories and public datasets. These materials make the disaster a valuable case for spatial analysis of long-term industrial risk.

Geospatial analysis can be used not only to reconstruct catastrophic events but also to analyze persistent exposure, vulnerability, and intergenerational impacts. Buffering, exposure zoning, health overlays, and land-use change analysis allow researchers to trace how technological hazards unfold spatially and temporally (Cutter et al., 2008; Dhara & Dhara, 2002). Together, these applications demonstrate that technological disasters are not confined to a single moment in time. They unfold across space and generations, making them especially well-suited to geospatial analysis aimed at prevention rather than retrospective explanation.

6.3 Modeling the Fire and Tech Hazard Risk in Geospatial Terms

6.3.1 The Fire and Tech Hazard Threat is a Complex Intersection

Modeling the fire and technological hazards in cities requires translating complicated social, environmental, and infrastructural conditions into spatially explicit representations of risk. Unlike hazards such as earthquakes or floods, which are often modeled using relatively stable physical parameters, fire risk emerges from the interaction of human behavior, land use, materials, infrastructure, and environmental conditions that vary across space and time (Cutter et al., 2008; UNDRR, 2022). As a result, fire risk is best understood not as a single surface or probability but as a composite of overlapping risk factors.

Geospatial modeling provides a means of integrating these factors into a coherent analytical framework. By combining data on fuel sources, ignition likelihood, population exposure, and emergency response capacity, GIS enables analysts to identify where fires are most likely to occur, where they are most likely to spread, and where their impacts would be most severe (Burby, 2006). Importantly, these models support a shift from reactive response toward anticipatory planning—allowing cities to prioritize inspections, target mitigation investments, and design evacuation and response strategies before disasters unfold.

6.3.2 Conceptual Tools for Fire Risk Modeling

Fire risk modeling relies on a set of conceptual tools that frame how hazards and vulnerability are represented spatially. At its core, most approaches draw on a risk triangle or risk equation, in which overall risk is understood as a function of hazard, exposure, and vulnerability (Wisner et al., 2004). GIS enables each of these components to be mapped, analyzed, and combined explicitly.

Hazard layers represent the likelihood and potential intensity of fire or technological failure. These may include vegetation type and fuel load, land surface temperature, historical fire occurrence, locations of hazardous industrial facilities, and meteorological variables such as wind speed and direction (Radeloff et al., 2005; IPCC, 2022). In urban contexts, hazard layers often incorporate building materials, electrical infrastructure, and fuel storage sites that increase ignition potential.

Exposure layers capture what is at risk. Population density, housing types, building footprints, workplaces, schools, and critical infrastructure define where people and assets intersect with potential fire hazards. High exposure does not imply inevitable loss, but it magnifies the consequences of failure when fires occur (Cutter et al., 2008).

Vulnerability layers reflect differential capacity to anticipate, cope with, and recover from fire events. Indicators may include income, age structure, disability, housing informality, access to transportation, and proximity to emergency services. Vulnerability is deeply shaped by political and economic conditions, making it uneven across urban space (UN-Habitat, 2020).

Together, these layers support composite fire risk surfaces, hotspot analyses, and scenario modeling that allow planners to explore how changes in land use, infrastructure, or climate conditions alter urban fire risk. These conceptual tools form the analytical foundation for the applied GIS methods introduced in the sections that follow.

6.3.3 Project Design Questions

Effective fire risk modeling begins with project design questions that link spatial analysis to decision-making. Rather than asking whether a city is "at risk," analysts must specify where, from what, and for whom fire and technological hazards pose the greatest threat (Cutter et al., 2008). Key questions include: Where are ignition sources most concentrated? Which neighborhoods combine high fuel loads with dense populations? How quickly can emergency services reach these areas under normal and extreme conditions? What populations face barriers to evacuation or recovery?

Equally important are governance-oriented questions. Which hazards are regulated, and which operate informally? Where do zoning rules allow hazardous activities near residential areas? How might climate variability alter fire behavior or infrastructure reliability over time (IPCC, 2022)? Framing these questions spatially ensures that GIS analysis supports prevention, preparedness, and targeted intervention, if possible, rather than post-disaster documentation alone.

6.3.4 Lab: Mapping the Wildland–Urban Interface (WUI) and Fire Risk in Los Angeles

This lab introduces a practical geospatial workflow for identifying people and property at risk from wildfires at the WUI. The objective is not to predict individual fire events but to demonstrate how spatial definitions of wildlands, ignition zones, and ember exposure can be combined with population and property data to reveal patterns of vulnerability in a large metropolitan region.

Los Angeles provides a compelling case study. The city's expansion into surrounding hills and chaparral landscapes has produced extensive WUI conditions, where dense residential development intersects with highly fire-prone vegetation. Similar patterns are increasingly common in cities worldwide as urban growth pushes into forests, savannas, and grasslands (Radeloff et al., 2005; Syphard et al., 2017).

Learning Objectives

By the end of this lab, students should be able to:

- Define and map key components of the WUI
- Identify populations and assets exposed to wildfire risk
- Use overlay and zonal analysis to classify "at-risk" areas
- Translate fire risk mapping into planning-relevant insights

6.3.4.1 Step 1: Define the Study Area

Begin by defining a clear study boundary. In this example, the study area is Los Angeles County, which contains extensive WUI conditions across its northern, western, and eastern edges.

Administrative boundary data can be obtained from county or state GIS portals, the Humanitarian Data Exchange, or OpenStreetMap-derived products (United Nations Office for the Coordination of Humanitarian Affairs [OCHA], 2023). Clipping all subsequent layers to a common boundary ensures consistency across datasets and avoids edge effects in spatial analysis.

6.3.4.2 Step 2: Assemble Wildland–Urban Interface Layers

The WUI can be conceptualized as a set of nested zones extending outward from wildland vegetation into developed areas.

6.3.4.2.1 Wildland Areas

Wildlands are defined as large, predominantly vegetated patches capable of sustaining wildfire. While definitions vary, many studies suggest a minimum contiguous area of approximately two square miles (about 5 km^2) to meaningfully represent wildfire fuel landscapes (Radeloff et al., 2005; Syphard et al., 2017).

In the United States, wildland and WUI layers are available through FEMA's geospatial portal and the U.S. Fire Administration's WUI datasets. For global applications, comparable datasets are available through Esri's Global Wildland–Urban Interface layer hosted on ArcGIS Online, which requires email registration (FEMA, 2023; Esri, 2022).

6.3.4.2.2 Ignition Zones

Ignition zones represent areas immediately adjacent to wildlands where fires are most likely to start or spread into developed areas. A commonly used threshold is approximately 100 feet (about 30 meters) from the wildland boundary, reflecting zones of direct flame contact and human-caused ignitions (US Fire Administration, 2021).

If pre-curated ignition zones are unavailable, they can be generated by buffering wildland polygons outward by the specified distance.

6.4.2.2.3 Ember Zones

Ember zones capture areas vulnerable to wind-driven embers, which can ignite structures well beyond the active fire front. Empirical studies show embers can travel up to 1.5 miles (about 2.4 km) under extreme wind conditions, making ember exposure one of the dominant causes of structural loss in WUI fires (Radeloff et al., 2005; Syphard et al., 2017).

Ignition zones fall entirely within ember zones, creating a nested risk structure:

$$\text{Wildlands} \rightarrow \text{Ignition Zone} \rightarrow \text{Ember Zone}$$

Merging these layers produces a composite WUI exposure surface suitable for analysis.

6.3.4.3 Step 3: Add Population and Property Data

To move from hazard mapping to exposure analysis, demographic and asset data can be integrated.

6.3.4.3.1 Population Data

For the United States, population counts can be obtained from the U.S. Census Bureau at multiple scales, including census tracts, block groups, or ZIP Code Tabulation Areas. The choice of scale should reflect planning needs and data availability.

For applications outside the United States, gridded population products such as WorldPop provide comparable estimates, particularly in regions where census data are outdated or incomplete (WorldPop, 2023).

6.3.4.3.2 Property Value Data

Property data highlight the economic stakes of wildfire exposure.

In this lab, property values are drawn from the Zillow Home Value Index (ZHVI), aggregated by ZIP (postal) code. These data provide smoothed, seasonally adjusted estimates of typical home values and are commonly used in spatial analyses of housing exposure to environmental risk (Zillow, 2023).

6.3.4.4 Step 4: Identify "At-Risk" ZIP Codes

With WUI, population, and property layers prepared, identify areas most exposed to wildfire risk.

1. Merge the wildland, ignition, and ember zones into a single WUI mask.

2. Perform a tabulate intersection or equivalent zonal analysis between ZIP codes and the WUI mask.
3. Calculate the percentage of each ZIP code overlapping WUI areas.
4. Classify ZIP codes as "at risk" if 50 percent or more of their area falls within wildland, ignition, or ember zones.

This threshold is adjustable and should be explicitly justified based on planning or policy objectives (Cutter et al., 2008).

6.3.4.5 Step 5: Visualize Risk

Two core visualizations help translate analysis into decision-relevant outputs:

1. At-Risk ZIP Codes by Property Value
   Mapping typical home values within WUI-exposed ZIP codes illustrates how wildfire risk is often concentrated in high-value housing at the urban fringe.
2. At-Risk ZIP Codes by Population
   Mapping population counts highlights where wildfire exposure intersects with evacuation demand and emergency response capacity.

Terrain or hillshade layers may be added to contextualize slope and fire behavior, which strongly influence fire spread in WUI settings (Radeloff et al., 2005).

Figure 6.4 is a diagram depicting the WUI, to help with interpreting the two Urban-Wilderness Interface maps for Los Angeles County, California.

Figure 6.5 shows areas denoted as wildlands, ignition zones, and ember zones.

Figure 6.6 shows areas with at-risk populations, and Figure 6.7 depicts areas with the most dollar-valued property at risk. Figures 6.4–6.7 are all credited to Maya O'Brien.

6.3.4.6 Step 6: Interpretation and Extensions

This analysis demonstrates that wildfire risk at the urban edge is not evenly distributed. In Los Angeles, large populations and substantial property value lie within ember exposure zones, underscoring the consequences of long-term land-use decisions that place dense development in fire-prone landscapes.

Extensions of this lab may include evacuation modeling, response-time analysis, or comparative studies across regions. The same workflow can be adapted to cities outside the United States using global WUI, land-cover, and population datasets (UNDRR, 2022).

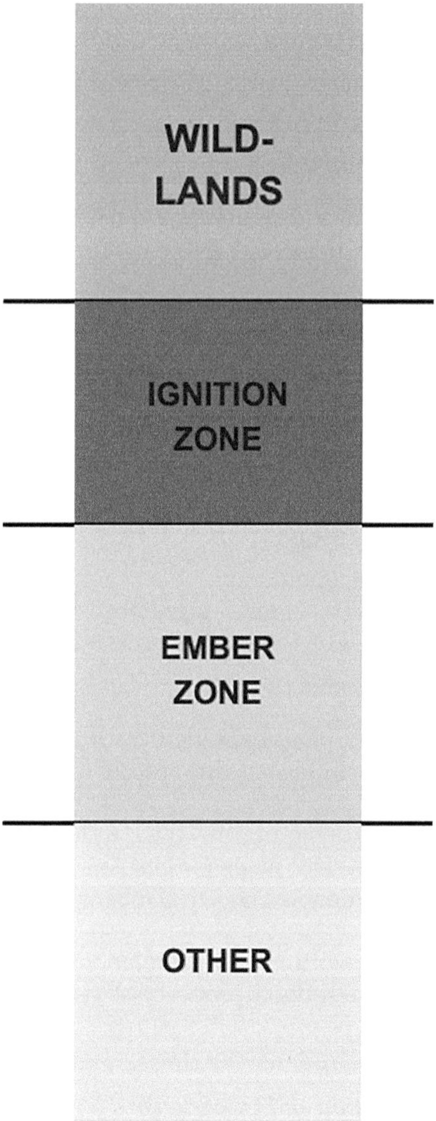

FIGURE 6.4
Key for urban-wilderness interface zone maps.

6.3.4.6.1 Key Takeaway

The WUI is not a thin boundary, but a graduated risk zone shaped by variables such as vegetation, wind, topography, and development patterns. GIS provides a powerful framework for making these risks visible, supporting

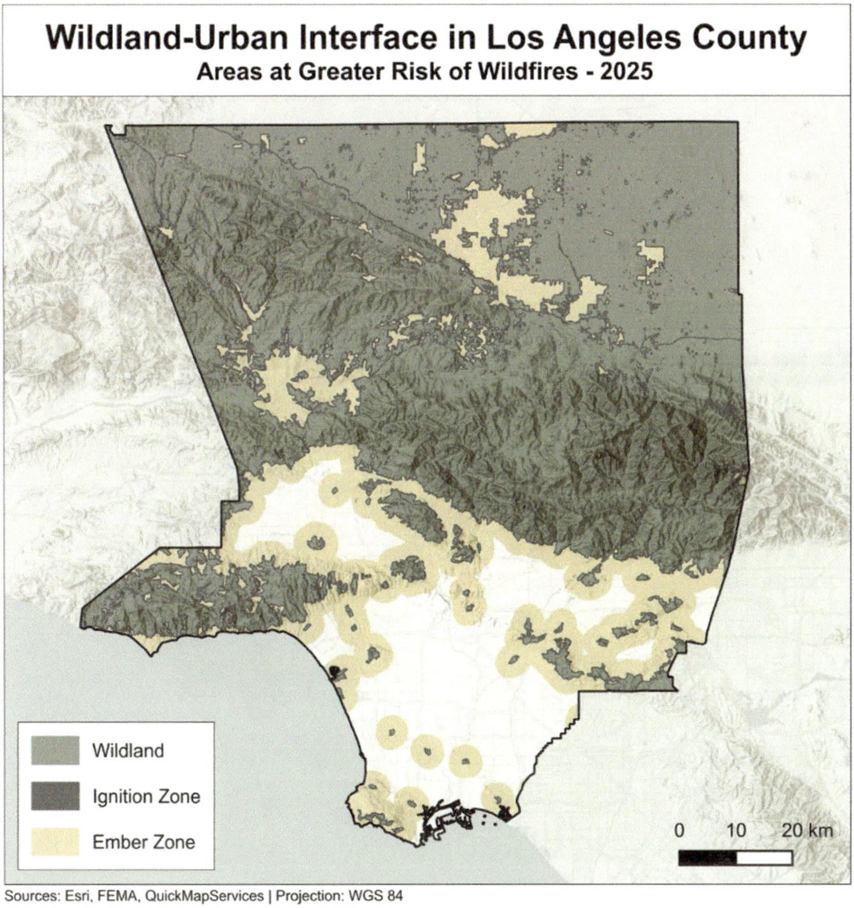

FIGURE 6.5
Urban-wilderness interface zone map: Areas at greatest risk.

land-use planning and risk reduction strategies before fires occur rather than after losses are incurred (Cutter et al., 2008).

6.3.5 Advanced Modeling Approaches

Beyond basic overlay analysis, advanced geospatial methods allow cities to explore how fires and technological hazards behave under dynamic conditions. These approaches are particularly valuable for preparedness planning and stress-testing response systems.

Network and response-time modeling evaluates how quickly emergency services can reach high-risk areas. By combining road networks, traffic

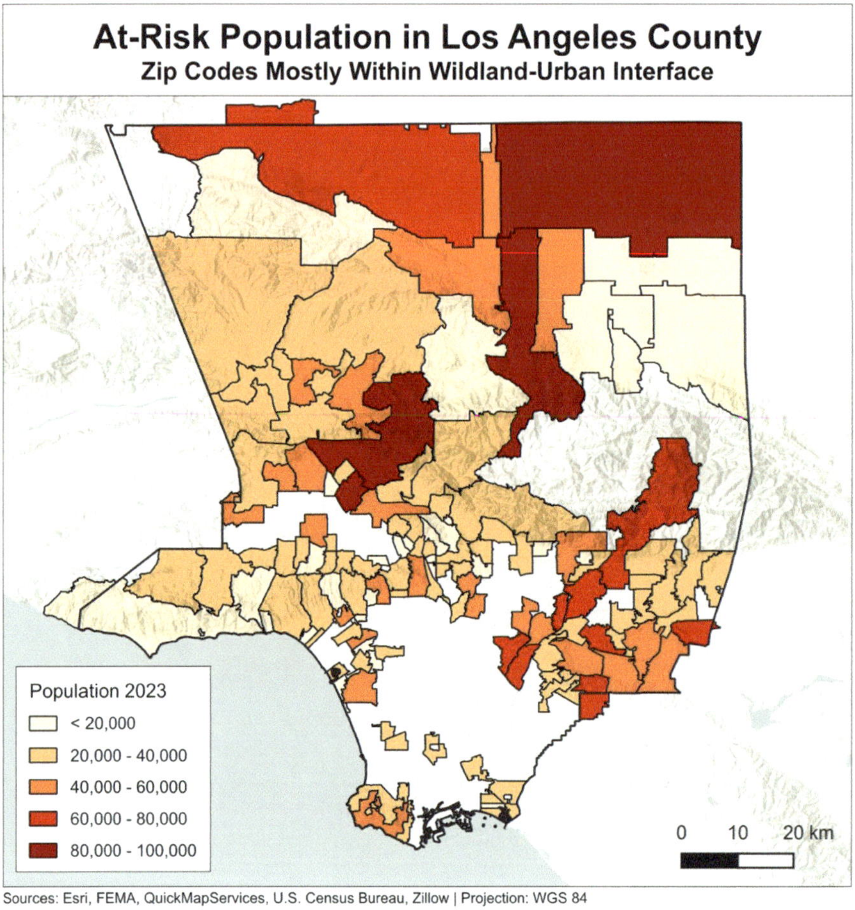

FIGURE 6.6
Urban-wilderness interface zone map: At-risk population in LA county.

conditions, station locations, and travel-time thresholds, GIS can identify neighborhoods where delayed response significantly increases fatality risk. Such analyses often reveal sharp inequalities between well-served and marginalized areas (OECD, 2018).

Fire spread and scenario modeling incorporates wind, slope, fuel continuity, and built form to simulate how fires may evolve under different conditions. While detailed physics-based models are data-intensive, simplified scenario approaches—such as buffer-based ember zones or directional spread models—can still inform evacuation planning and land-use decisions (Radeloff et al., 2005; Syphard et al., 2017).

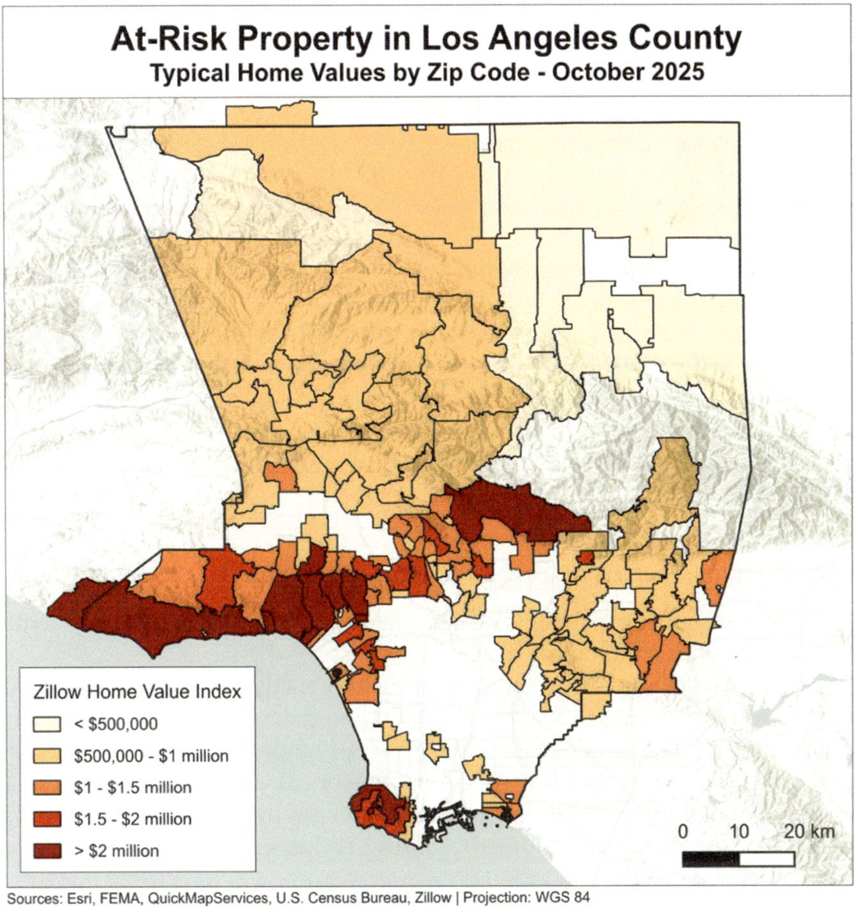

FIGURE 6.7
Urban-wilderness interface zone map: Home values in at-risk UWI zone.

Uncertainty modeling is critical in fire risk analysis. Data gaps, informal development, and rapidly changing land use introduce uncertainty that should be acknowledged rather than ignored. Sensitivity testing, which means adjusting weights or assumptions to observe how risk patterns change, helps planners distinguish robust findings from artifacts of data choice (Cutter et al., 2008).

Finally, and most importantly, **scenario analysis** allows cities to ask "what if" questions: What happens to fire risk if a new industrial zone is built? If climate-driven heat waves become more frequent? If emergency response capacity improves or degrades? By embedding these scenarios in GIS, planners can evaluate policy choices spatially before they are implemented.

6.4 Getting to Action on Fire and Technological Threats

Understanding fire and technological hazards is only the first step. Reducing risk depends on translating spatial insight into interventions that reshape land use, infrastructure, and governance. Fire prevention is not primarily a technical challenge, but an institutional and planning problem that plays out unevenly across urban space.

Geospatial analysis is particularly powerful in this domain because it reveals where hazards, people, and institutional capacity intersect. Fire risk maps do more than document danger. They identify priority zones for inspection, enforcement, investment, and community engagement, enabling cities to act before an ignition occurs rather than responding after a loss.

This section outlines practical pathways for action, emphasizing interventions that are spatially targeted, institutionally feasible, and scalable across cities, particularly in the Global South.

6.4.1 Principles for Enhancing Urban Fire Resilience

Several cross-cutting principles emerge from the global experience with urban fires and technological disasters.

First, prevention is more effective and less costly than response. Most urban and industrial fires are preventable through basic measures, including safe electrical systems, separation of hazardous land uses, limits on combustible materials, and routine inspections. Historical declines in fire mortality in Europe and North America were driven not by improved firefighting alone, but by decades of enforced building codes, zoning regulations, and electrical standards.

Second, risk is spatial rather than random. Fire hazards cluster where fuel loads, ignition sources, and vulnerable populations coincide. Informal settlements with dense housing, industrial districts storing hazardous materials, and WUIs all exhibit predictable patterns of exposure. Mapping these patterns allows cities to prioritize limited resources rather than applying uniform solutions that fail to address underlying risk.

Third, enforcement matters as much as regulation. Many cities possess fire codes and safety standards on paper but lack the inspection capacity, political backing, or institutional incentives needed to enforce them consistently. GIS-based inspection planning, which targets known high-risk zones, can significantly increase the effectiveness of limited enforcement staff.

Fourth, resilience is social as well as technical. Community awareness, trust in authorities, and local response capacity strongly shape outcomes during fire events. Neighborhoods with organized residents, clear communication channels, and basic preparedness tend to suffer fewer fatalities even when fires occur.

Finally, fire safety must be embedded in land-use planning. Once hazardous facilities or flammable housing are embedded within dense urban fabrics, retrofitting becomes expensive and politically difficult. The most effective interventions occur upstream, before land is developed, permits are issued, or materials are approved.

Together, these principles emphasize that fire resilience is not achieved through isolated projects but through sustained alignment between spatial planning, infrastructure investment, and governance.

6.4.2 Pathways to Action

Several practical pathways translate spatial analysis into risk reduction.

Cities should begin by mapping and monitoring hazard zones. Dynamic, publicly accessible maps that integrate vegetation and fuel loads, industrial facilities, building materials, population density, and historical incident data allow authorities to anticipate periods of elevated risk. Regular updates are particularly important during droughts, heat waves, and other high-risk conditions.

Land-use separation remains one of the most powerful fire prevention tools. High-risk industrial activities, fuel depots, and hazardous storage sites should be spatially separated from dense residential areas wherever possible. Where separation is not feasible, buffer zones, structural hardening, and enhanced inspection regimes become essential.

Emergency response capacity can also be strengthened through spatial analysis. GIS-based network modeling helps identify neighborhoods where fire response times exceed critical thresholds. These insights support strategic placement of fire stations, upgrades to water infrastructure, and prepositioning of equipment during high-risk periods.

Transparency and accountability play an equally important role. Publishing inspection results, violation notices, and remediation status builds public trust and creates pressure for compliance. Open data portals allow journalists, researchers, and community organizations to scrutinize safety practices, shifting fire prevention from closed bureaucratic processes toward shared civic responsibility.

Finally, cities can build community-based fire capacity. In many Global South cities, formal emergency services cannot reach all neighborhoods quickly. Community fire brigades, early-warning systems, and mobile reporting platforms provide critical first-response capacity and improve situational awareness when professional responders are delayed.

6.4.3 Mobile Inspection and Reporting Technologies

Recent advances in mobile technology offer practical tools for strengthening fire safety enforcement and community engagement. Digital inspection applications allow inspectors to collect standardized, geotagged data on building conditions, electrical systems, fuel storage, occupancy levels, and code compliance. Commonly recorded variables include building use, construction materials, electrical load, presence of fire exits and alarms, proximity to hazardous facilities, and remediation status. These systems reduce paperwork, improve consistency, and generate spatial datasets that can be analyzed for systemic risk patterns.

Several cities have piloted mobile inspection platforms with promising results. In parts of southern Africa, digital tools have been used to track compliance in informal and industrial buildings, reducing opportunities for bribery and data loss. In Freetown, Sierra Leone, mobile inspection and reporting systems have supported broader disaster risk reduction efforts by linking field observations directly to planning agencies.

Community-facing reporting tools are equally important. Allowing residents to report blocked exits, illegal fuel storage, or electrical hazards through simple mobile interfaces increases situational awareness and expands the reach of limited enforcement agencies. When integrated with GIS, these reports help authorities identify emerging risks before disasters occur.

The key insight is that technology does not replace regulation. It amplifies institutional capacity by extending reach, improving transparency, and generating the spatial intelligence needed for targeted intervention.

6.4.4 Human and Infrastructure Impacts

Fires and technological disasters damage far more than individual buildings. Electrical grids, water systems, transportation networks, and communication infrastructure are often co-located with high-risk industrial and residential zones. Fires can disable substations, rupture pipelines, contaminate water supplies, and disrupt access routes, multiplying indirect losses long after the flames are extinguished. Mapping infrastructure alongside fire risk reveals cascading vulnerabilities in which damage to a single facility can cripple entire districts.

One human impact deserving particular attention is toxic exposure from modern materials. Contemporary urban fires increasingly involve synthetic materials such as plastics, foams, laminates, and treated textiles that burn at high temperatures and release highly toxic smoke. In many residential and industrial fires, smoke inhalation rather than burns is the leading cause of death.

Burning plastics release dioxins, furans, and other persistent pollutants that contaminate air, soil, and water. These toxins pose long-term health risks, particularly for children and first responders, and can render sites hazardous long after rebuilding begins. Reducing reliance on highly combustible synthetic materials through building codes, taxation, or bans represents a powerful but underutilized fire prevention strategy.

6.4.5 Looking Ahead

Urban fire and technological hazards reveal a fundamental tension in contemporary cities. The same systems that enable economic growth, including dense housing, industrial concentration, and energy-intensive infrastructure, also generate acute and chronic risks when poorly governed.

The evidence is clear. Fires are not inevitable. They follow predictable spatial patterns shaped by land use, materials, infrastructure, and inequality. Where cities invest in planning, enforcement, and transparency, fire mortality declines dramatically. Where they do not, disasters repeat.

Looking ahead, several priorities stand out. Fire risk must be integrated explicitly into land-use planning rather than treated solely as an emergency services problem. Climate variability will continue to amplify risk, making proactive governance increasingly urgent. Technological tools, from satellite fire detection to mobile inspection platforms, offer unprecedented capacity to monitor danger, but only if institutions are empowered to act on the data they generate.

At a broader level, reducing fire risk also requires rethinking material and energy choices. Limiting flammable construction materials, discouraging toxic consumer products, and designing cities that rely less on dangerous fuels and overloaded electrical systems align fire prevention with sustainability goals.

The evidence is clear. Fires are not inevitable. They follow predictable spatial patterns shaped by land use, materials, infrastructure, and inequality. Where cities invest in planning, enforcement, and transparency, fire mortality declines dramatically. Where they do not, disasters repeat.

The task ahead is therefore not to invent new tools, but to use existing spatial knowledge to guide decisions before risk becomes tragedy.

References

Ahmed, F. E., Islam, M. S., & Hasan, M. (2014). Fire safety in the readymade garment sector of Bangladesh: A critical analysis. *International Journal of Occupational Safety and Ergonomics, 20*(3), 463–474. https://doi.org/10.1080/10803548.2014.11077055

Alamgir, F., & Alakavuklar, O. N. (2020). Compliance code and worker resistance in the Bangladesh garment industry: The Rana Plaza disaster and its aftermath. *Journal of Business Ethics, 167*, 459–477. https://doi.org/10.1007/s10551-019-04189-5

Alamgir, F., & Banerjee, S. B. (2019). Contested compliance regimes in global production networks: Insights from the Bangladesh garment industry. *Human Relations, 72*(2), 272–297. https://doi.org/10.1080/0969160X.2020.1795453

Bonjour, S., Adair-Rohani, H., Wolf, J., Bruce, N. G., Mehta, S., Prüss-Ustün, A., … Smith, K. R. (2013). Solid fuel use for household cooking: Country and regional estimates for 1980–2010. *Environmental Health Perspectives, 121*(7), 784–790. https://doi.org/10.1289/ehp.1205987

Bharadwaj, P., Lundborg, P., & Rooth, D.-O. (2014). Birth weight in the long run: Evidence from the Bhopal gas tragedy. *BMJ Open*, *4*(7), e004546.

Burby, R. J. (2006). Hurricane Katrina and the paradoxes of government disaster policy: Bringing about wise governmental decisions for hazardous areas. *Annals of the American Academy of Political and Social Science, 604*(1), 171–191. https://doi.org/10.1177/0002716205284676

Cutter, S. L., Barnes, L., Berry, M., Burton, C., Evans, E., Tate, E., & Webb, J. (2008). A place-based model for understanding community resilience to natural disasters. *Global Environmental Change, 18*(4), 598–606. https://doi.org/10.1016/j.gloenvcha.2008.07.013

CTIF (International Association of Fire and Rescue Services). (2021). *World fire statistics report no. 26*. CTIF Centre of Fire Statistics.

Dhara, V. R., & Dhara, R. (2002). The Union Carbide disaster in Bhopal: A review of health effects. *Archives of Environmental Health, 57*(5), 391–404. https://doi.org/10.1080/00039890209602941

Donaghey, J., & Reinecke, J. (2018). When industrial democracy meets corporate social responsibility—A comparison of the Bangladesh Accord and Alliance as responses to the Rana Plaza disaster. *British Journal of Industrial Relations, 56*(1), 14–42. https://doi.org/10.1111/bjir.12242

Esri. (2022). *Global Wildland–Urban Interface (WUI) layer*. ArcGIS Online. https://www.arcgis.com

Federal Emergency Management Agency (FEMA). (2023). FEMA Geospatial Resource Center (FGRC): Hazard and risk datasets. https://gis.fema.gov

Hackitt, J. (2018). *Building a safer future: Independent review of building regulations and fire safety*. UK Ministry of Housing, Communities & Local Government.

ILO (International Labour Organization). (2013). *Bangladesh: Fire safety in the ready-made garment sector*. ILO.

ILO. (2018). *Improving fire and building safety in the Bangladesh garment industry*. International Labour Organization.

IPCC. (2022). *Climate change 2022: Impacts, adaptation, and vulnerability (AR6 WGII)*. Cambridge University Press.

Landrigan, P. J., Fuller, R., Acosta, N. J. R., Adeyi, O., Arnold, R., Baldé, A. B., … Zhong, M. (2018). The Lancet Commission on pollution and health. *The Lancet, 391*(10119), 462–512. https://doi.org/10.1016/S0140-6736(17)32345-0

Lane, B. (2020). The Grenfell Tower fire: Lessons for fire safety regulation. *Fire Safety Journal, 113*, 102937.

National Weather Service. (2023). *Maui wildfire event summary: August 8–9, 2023*. NOAA National Weather Service, Honolulu Weather Forecast Office. https://www.weather.gov/hfo/MauiFires

NFPA (National Fire Protection Association). (2019). *Fire loss in the United States*. NFPA Research.

OECD. (2018). *Rebuilding urban resilience: Cities and disaster risk reduction*. OECD Publishing.

Radeloff, V. C., Hammer, R. B., Stewart, S. I., Fried, J. S., Holcomb, S. S., & McKeefry, J. F. (2005). The wildland–urban interface in the United States. *Ecological Applications, 15*(3), 799–805. https://doi.org/10.1890/04-1413

Reinecke, J., & Donaghey, J. (2015). After Rana Plaza: Building coalitional power for labour rights between unions and (consumption-based) social movement organisations. *Organization, 22*(5), 720–740. https://doi.org/10.1177/1350508415585028

State of Hawai'i Office of the Governor. (2023). Maui wildfires disaster response and recovery updates. https://governor.hawaii.gov/newsroom/maui-wildfires/

Syphard, A. D., Keeley, J. E., Pfaff, A. H., & Ferschweiler, K. (2017). Human presence diminishes the importance of climate in driving fire activity across the United States. *Proceedings of the National Academy of Sciences, 114*(52), 13750–13755. https://doi.org/10.1073/pnas.1713885114

UNDRR (United Nations Office for Disaster Risk Reduction). (2022). *Global assessment report on disaster risk reduction*. United Nations.

UN-Habitat. (2020). *World cities report 2020: The value of sustainable urbanization*. United Nations Human Settlements Programme.

United Nations Office for the Coordination of Humanitarian Affairs (OCHA). (2023). *Humanitarian Data Exchange (HDX)*. https://data.humdata.org

U.S. Fire Administration. (2021). *Wildland–urban interface fire risk and mitigation guidance*. Federal Emergency Management Agency.

Varma, D. R., & Varma, R. (2005). The Bhopal disaster of 1984. Council of Canadians.

Wisner, B., Blaikie, P., Cannon, T., & Davis, I. (2004). *At risk: Natural hazards, people's vulnerability and disasters* (3rd ed.). Routledge.

WHO (World Health Organization). (2016). *Burn prevention: Success stories and lessons learned*. WHO Press.

WorldPop. (2023). *WorldPop global population data*. University of Southampton. https://www.worldpop.org

Yin, X., Huang, S., Zhu, Z., Ma, Q., Wang, Y., Liu, X., Shen, T., & Zhu, F. (2025). The global, regional, and national burden of burns: An analysis of injury by fire, heat, and hot substances in the global burden of disease study 2019. Injury, 56, 111955. https://doi.org/10.1016/j.injury.2024.111955

Zillow. (2023). Zillow Home Value Index (ZHVI). Zillow Research. https://www.zillow.com/research/data/

7

Mapping Heat and Urban Heat Islands

Chapter Introduction

Heat is a silent hazard. Unlike floods or fires, heat builds slowly and kills quietly. In cities, extreme heat thrives in dense urban zones with limited vegetation and impervious surfaces that trap and re radiate solar energy. As climate change intensifies, heat has become one of the most pervasive and underestimated environmental threats facing urban populations worldwide.

Urban heat does not affect all residents equally. Neighborhoods with sparse tree cover, informal housing, poor ventilation, and limited access to cooling infrastructure experience disproportionately higher exposure. At the level of the individual and household, heat interacts with poverty, age, health status, and labor patterns to produce uneven risk, making it both an environmental and social hazard. This chapter examines how heat operates as an urban threat, how it can be modeled geospatially, and how spatial analysis can guide interventions that reduce exposure and save lives.

This chapter opens with a case study of the extreme heat event that struck New Delhi in 2024, illustrating heat as a slow-moving but deadly urban hazard. It then unpacks the science of heat and urban heat islands, demonstrates how heat can be mapped using freely available geospatial data, and concludes with strategies for translating analysis into action.

7.1 Setting the Scene: Living at the Edge of Thermal Tolerance

In May and June 2024, northern India experienced one of the most intense and spatially extensive heat waves ever recorded in the instrumental era. In New Delhi, daytime air temperatures repeatedly exceeded 45°C (113°F), while nighttime temperatures remained above 32°C (90°F) for prolonged periods. These conditions pushed the city's energy and public health systems to their limits. Electricity demand surged as residents sought mechanical cooling, triggering rolling blackouts that disproportionately affected informal settlements and older neighborhoods with weak grid connections. Hospitals reported sharp increases in heat exhaustion, dehydration, and cardiovascular distress,

DOI: 10.1201/9781032636948-7

while schools closed and construction activity slowed or halted entirely. What made this event especially dangerous was not simply the magnitude of peak temperatures, but their persistence. Coinciding with elevated humidity, high temperatures acted on the dense, heat-retaining built environment of a megacity already under chronic environmental stress (India Meteorological Department, 2024; World Health Organization, 2023).

Heat waves are often framed as episodic disasters, short-lived anomalies that briefly cause spikes in mortality before conditions return to normal. Increasingly, however, extreme heat is better understood as a chronic urban condition that exposes structural vulnerabilities in both infrastructure and human physiology. Unlike floods or cyclones, which produce visible damage and trigger formal emergency declarations, heat accumulates silently in bodies, buildings, and crowded streets. Its impacts fall disproportionately on populations whose suffering is easily overlooked: older adults, infants, outdoor workers, people with chronic illnesses, and the socially isolated. As a result, heat-related mortality is routinely undercounted, with deaths frequently attributed to cardiovascular or respiratory causes without acknowledging heat as a contributing factor. Excess-mortality studies consistently show that extreme heat now rivals or exceeds other weather-related hazards as a cause of premature death in many regions (Gasparrini et al., 2015; WHO, 2023).

The danger posed by extreme heat is also profoundly uneven across urban space. Within a single city, air and surface temperatures can vary by several degrees depending on land cover, building materials, street geometry, and access to vegetation. Neighborhoods dominated by asphalt, concrete, and corrugated metal absorb large amounts of solar radiation during the day and release it slowly at night, producing elevated minimum temperatures that prevent physiological recovery. In contrast, areas with tree canopy, permeable surfaces, and proximity to water bodies experience lower daytime temperatures and more rapid nocturnal cooling. These contrasts—often grouped under the term *urban heat island*—translate into unequal health outcomes and are associated with factors such as income, tenure, and historical patterns of disinvestment (Oke, 1982; Voogt & Oke, 2003; Harlan et al., 2013). Heat, in this sense, is not merely an environmental exposure but a spatial expression of social inequality.

At the global scale, the geography of heat risk is undergoing a rapid transformation. For the most of human history, extreme heat was geographically bounded, concentrating in deserts and tropical lowlands where building forms and cultural practices evolved in partial accommodation to thermal stress. Climate change is now pushing dangerous heat into regions that lack both long-term acclimatization and adaptive infrastructure. Case in point: The European heat wave of 2003, associated with an estimated 70,000 excess deaths, remains a canonical example. Many victims lived alone in urban apartments that lacked air conditioning and were socially isolated, revealing vulnerability patterns like inequitable access to cooling and strained electrical grids (Sera et al., 2019) that were at the time more commonly associated with cities in the Global South (Vandentorren et al., 2006; Robine et al., 2008). Two decades later, similar patterns are emerging across the planet, even in northern Europe, Canada, and Russia.

What distinguishes the current moment from heat waves of the past is the convergence of three forces: rising baseline temperatures driven by anthropogenic climate change; rapid urbanization that replaces vegetated landscapes with heat-retaining surfaces; and growing socioeconomic inequality that constrains adaptive capacity for large segments of the urban population. Together, these forces compress the margin between survivable heat and physiological failure. Scientists increasingly describe this margin using *wet-bulb temperature*, a combined measure of heat and humidity that reflects the body's capacity to cool itself through sweating. Sustained wet-bulb temperatures above approximately 35°C are widely considered incompatible with human survival, even for young, healthy individuals resting in the shade. Recent heat events in South Asia and the Persian Gulf have approached this wet-bulb threshold, and localized urban microclimates may already have exceeded it briefly (Sherwood & Huber, 2010; Raymond et al., 2020).

This chapter treats extreme heat not simply as a meteorological phenomenon, but as a spatially structured urban threat that can be modeled using geospatial tools. Maps of land-surface temperature, vegetation cover, building density, and population vulnerability provide insights into where heat risk is greatest and why. At the same time, heat is not an abstract raster layer but is experienced bodily, shaped by housing quality, labor conditions, access to electricity, and social networks. The task of heat mapping, therefore, is not only to identify hotspots but also to connect physical exposure to lived experience and to the planning, policy, and design decisions that render some neighborhoods far more dangerous than others.

The sections that follow unpack the science of extreme heat and urban heat islands. It will then examine how thermal exposure and vulnerability are represented in spatial data and demonstrate practical workflows for analyzing heat risk using freely available datasets. This chapter concludes by considering adaptation strategies from urban greening and cooling centers to building design and the ethical questions raised by a future in which some cities may periodically approach, or exceed, the limits of human habitability.

7.2 Unpacking the Threat: How and Why Heat Becomes Deadly in Cities

Extreme heat is not merely an atmospheric condition but a spatially produced urban phenomenon. Cities systematically modify surface energy balances through building materials, land cover, and urban form, creating localized

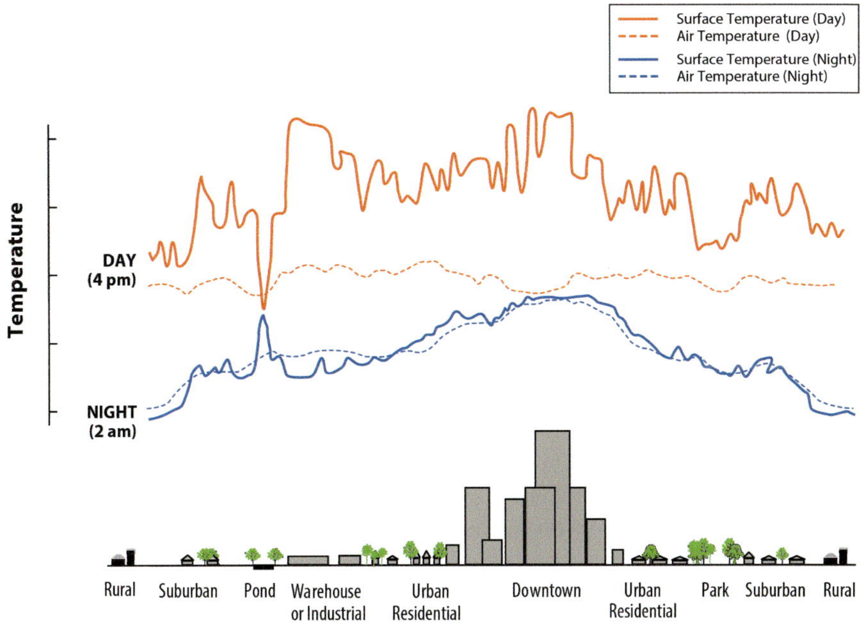

FIGURE 7.1
Urban heat island diagram. (https://www.epa.gov/heatislands/what-are-heat-islands. Public Domain.)

thermal environments that diverge sharply from surrounding rural areas. The resulting Urban Heat Island (UHI) effect, first described by Luke Howard in nineteenth-century London, remains one of the most well-documented examples of anthropogenic climate modification at the city scale (Howard, 1833; Oke, 1982). Figure 7.1 depicts two temperature lines against a rural to suburban to downtown skyline, depicting the urban heat island effect.

At its core, urban heat is governed by the surface energy balance, in which incoming shortwave solar radiation is partitioned into sensible heat, latent heat, ground heat flux, and reflected radiation. Urban surfaces such as asphalt, concrete, and many roofing materials typically have low albedo, high heat capacity, and limited moisture availability. These properties increase daytime heat absorption and slow nocturnal cooling, producing elevated nighttime temperatures that are particularly dangerous for human health (Oke, 1987; Voogt & Oke, 2003). Unlike rural landscapes, where evapotranspiration from vegetation provides an effective cooling mechanism, impervious urban surfaces suppress latent heat flux and redirect energy into sensible heat that directly warms the air.

Urban geometry further amplifies heat retention. Dense building clusters create "urban canyons" that trap longwave radiation emitted from surfaces after sunset, reducing radiative heat loss to the atmosphere.

This effect is especially pronounced in mid- and high-rise districts, where building height-to-street-width ratios limit sky-view factors and inhibit convective cooling (Arnfield, 2003). As a result, many cities experience their most dangerous thermal conditions not during peak daytime heating, but overnight, when buildings retain accumulated heat and residents lack physiological relief.

Surface roughness and aerodynamic properties also shape urban heat patterns. Increased roughness can enhance turbulent mixing and sometimes reduce daytime temperatures, but it may simultaneously limit nighttime ventilation under low-wind conditions. This is particularly true in informal settlements and tightly packed residential districts (Roth, 2007). The combined effect is a highly uneven urban thermal landscape. Heat manifests not as a uniform blanket over the city, but as a mosaic of microclimates structured by land use, building form, vegetation cover, and proximity to water bodies.

Vegetation plays a central countervailing role in urban thermal regulation. Tree canopy and other green infrastructure reduce surface and air temperatures through shading, evapotranspiration, and modification of local airflow. Empirical studies consistently show that neighborhoods with higher vegetation density experience lower daytime surface temperatures and reduced nighttime heat retention (Bowler et al., 2010; Ziter et al., 2019). These cooling benefits, however, are unevenly distributed within cities, often reflecting long-standing socioeconomic and racial inequities embedded in planning decisions (Hoffman et al., 2020).

Remote sensing has been instrumental in advancing scientific understanding of urban heat processes. Thermal infrared imagery from satellites such as Landsat enables the derivation of Land Surface Temperature (LST), providing spatially continuous measures of surface heating across entire metropolitan regions. While LST is not equivalent to near-surface air temperature, it is strongly correlated with localized heat stress and is particularly valuable for identifying persistent hotspots associated with impervious surfaces and limited vegetation (Weng, 2009).

Climate change is intensifying these mechanisms. Rising baseline temperatures increase the frequency, duration, and severity of heat waves, while urban expansion enlarges the spatial footprint of heat-amplifying surfaces. Attribution studies demonstrate that anthropogenic climate change has already increased the likelihood of extreme urban heat events across all continents, with particularly strong effects in rapidly growing metropolitan regions (Perkins-Kirkpatrick & Lewis, 2020; Raymond et al., 2020).

7.2.1 How Heat and Human Physiology Shape Health Risk

From a physiological perspective, heat becomes lethal when the human body can no longer maintain thermal equilibrium with its environment. Humans regulate core temperature primarily through sweating, a process of evaporative cooling that depends on temperature gradients and sufficient vapor pressure deficit. When the ambient air temperature approaches or exceeds the skin temperature, approximately 35°C, and the humidity is high, evaporative cooling becomes ineffective. Core body temperature rises, placing stress on the cardiovascular system and increasing the risk of heat exhaustion, heat stroke, and death (Parsons, 2014; Kenny et al., 2017).

Crucially, lethal heat does not require record-breaking daytime temperatures. Sustained exposure to moderately elevated heat, particularly when combined with warm nights, can be equally dangerous. Epidemiological studies consistently show that elevated nighttime minimum temperatures are strongly associated with excess mortality because they prevent physiological recovery following daytime exposure (Anderson & Bell, 2011; Vicedo-Cabrera et al., 2018).

Heat-related mortality is also systematically undercounted. Deaths are frequently attributed to cardiovascular or respiratory causes without acknowledging heat as a contributing factor. Excess-mortality analyses repeatedly demonstrate that heat waves produce substantial increases in deaths that are not captured by conventional reporting systems (Basu & Samet, 2002).

7.2.2 Urban Form, Housing, and Social Vulnerability

Urban heat risk is shaped not only by atmospheric conditions but also by housing quality and access to cooling. Buildings constructed with lightweight materials, metal roofing, poor insulation, and limited ventilation can reach indoor temperatures well above outdoor air temperature, effectively becoming heat traps (Tuholske et al., 2021). In many cities, informal housing lacks reliable electricity, making mechanical cooling unavailable or unaffordable.

Even in wealthier cities, reliance on air conditioning introduces new vulnerabilities. Power outages during heat waves can rapidly transform sealed, mechanically cooled buildings into dangerously hot enclosed spaces (Miller et al., 2023). Technological adaptation therefore does not eliminate heat risk; it redistributes it and, in some cases, amplifies it.

Labor conditions and socioeconomic inequality further intensify vulnerability. Outdoor and informal workers, including construction laborers, street vendors, and sanitation workers, experience prolonged exposure during the hottest hours of the day, often without legal protections or the ability to modify work schedules. Occupational heat stress is increasingly recognized as a major public health concern, particularly in low- and middle-income countries (Kjellstrom et al., 2018).

7.2.3 Climate Change and the Shrinking Margin of Safety

As with other urban environmental risks discussed in this book, climate change amplifies heat risk by increasing the frequency, duration, and intensity of heat waves while expanding their geographic reach. Of particular concern is the rise of high wet-bulb temperatures, which integrate heat and humidity into a single measure of physiological stress. Sustained wet-bulb temperatures above approximately 35°C are widely considered incompatible with human survival, even for young, healthy individuals resting in the shade (Sherwood & Huber, 2010).

Recent analyses show that such conditions are being approached with increasing frequency in parts of South Asia, the Persian Gulf, and coastal tropical regions, particularly within urban microclimates where humidity and heat retention are elevated (Raymond et al., 2020). These findings suggest that the margin between survivable heat and physiological failure is narrowing in many cities.

Heat exposure varies block by block and hour by hour, reflecting land cover, building density, vegetation, infrastructure, and social organization. Numerous studies demonstrate strong correlations between higher heat exposure and indicators of socioeconomic disadvantage, including income, race, tenure status, and historical patterns of segregation and disinvestment (Harlan et al., 2013; Shandas et al., 2019).

Understanding heat as an urban threat therefore requires a spatially explicit framework. Heat is not simply "hot weather" affecting a city uniformly. It is the outcome of decades of planning decisions, infrastructure investments, and governance choices that can be mapped, modeled, and modified. The next section translates these dynamics into geospatial terms, showing how temperature, land cover, and administrative data can be integrated to identify where heat risk is greatest and how it might be reduced.

7.3 Modeling Heat Risk in Geospatial Terms

Modeling urban heat as a spatial threat requires translating biophysical processes and social vulnerability into representations that can be mapped and used to guide action. Unlike hazards such as floods or landslides, heat leaves no permanent physical trace on the landscape. Its impacts are episodic, cumulative, and often invisible in conventional damage assessments. As a result, geospatial analysis of heat risk relies on proxy indicators—surface temperature, vegetation cover, building density, population characteristics, and access to infrastructure—that together approximate the conditions under which heat becomes dangerous.

The most widely used spatial representation of urban heat is LST, derived from thermal infrared remote sensing. Sensors aboard satellites such as Landsat measure emitted longwave radiation from the Earth's surface, which can be converted into surface temperature estimates using radiative transfer methods that account for atmospheric effects and surface emissivity (Weng, 2009; Voogt & Oke, 2003). While LST is not equivalent to near-surface air temperature or direct human exposure, it is strongly correlated with localized heat stress and is particularly valuable for revealing fine-scale spatial variability driven by land cover and built form.

LST datasets are typically raster-based, with pixel sizes ranging from 30 m (Landsat) to several kilometers (MODIS). Higher-resolution imagery allows analysts to identify neighborhood-scale hotspots associated with impervious surfaces, industrial zones, transport corridors, and areas lacking vegetation. When mapped repeatedly over time, LST can reveal lasting thermal trends rather than transient anomalies, highlighting places where residents are exposed to elevated heat year after year (Hoffman et al., 2020). However, careful interpretation is required. LST reflects surface conditions, not indoor temperatures or human exposure directly, and its accuracy can be affected by factors such as sensor overpass timing, cloud cover, and surface moisture conditions (Weng, 2009).

To contextualize surface temperature patterns, LST is often analyzed alongside vegetation indices, most commonly the Normalized Difference Vegetation Index (NDVI). NDVI serves as a proxy for canopy cover and photosynthetic activity, both of which are closely linked to evapotranspirative cooling (Bowler et al., 2010). Spatial overlays of LST and NDVI show inverse relationships, in that areas with higher vegetation density exhibit lower surface temperatures, while sparsely vegetated neighborhoods experience greater heat accumulation (Ziter et al., 2019). These relationships provide a mechanistic basis for linking urban greening interventions to measurable reductions in heat exposure.

Elevation and terrain also influence urban heat patterns, though their effects are often secondary to land cover in dense cities. Digital Elevation Models (DEMs) are used to assess slope, aspect, and airflow pathways that shape local microclimates. In coastal or hilly cities, elevation gradients may moderate nighttime cooling or exacerbate heat retention in low-lying basins where warm air pools under stable atmospheric conditions (Oke, 1987). Integrating DEM-derived variables into heat models helps distinguish topographic influences from anthropogenic drivers.

Heat exposure becomes a public health threat only when it intersects with vulnerable populations. Consequently, geospatial heat modeling increasingly incorporates socio-demographic data to move beyond physical temperature surfaces toward composite risk assessments. Population density grids, age distributions, income proxies, housing characteristics, and access to electricity or cooling infrastructure are commonly used to identify neighborhoods where heat exposure is likely to translate into morbidity and mortality (Harlan et al., 2013; Shandas et al., 2019). These datasets are often imperfect, particularly in rapidly growing cities where decennial census data lag behind urban expansion. Nonetheless, even coarse demographic indicators can reveal striking spatial disparities when overlaid with thermal patterns.

One common analytical approach is the construction of heat vulnerability indices, which combine multiple standardized variables, such as thermal exposure, sensitivity, and adaptive capacity, into a single composite surface (Reid et al., 2009). While such indices can be useful for prioritization and communication, they also carry methodological risks. Weighting schemes are often subjective, correlations among variables can obscure causal relationships, and composite scores may mask the specific mechanisms driving risk in particular locations. For this reason, scholars advocate using vulnerability indices as exploratory tools rather than definitive measures, emphasizing transparency and sensitivity analysis in their construction (Wolf et al., 2015).

Network-based analyses offer additional insight into the lived experience of heat. Accessibility modeling can be used to estimate walking distances or travel times to cooling centers or other public facilities during extreme heat events. Such analyses highlight how heat risk is shaped not only by exposure but also by mobility constraints, particularly for older adults, people with disabilities, and those with limited transport options (Adams et al., 2023). When combined with time-of-day considerations, network models can also reveal mismatches between service availability and periods of peak thermal stress.

Temporal analysis is essential for capturing the cumulative nature of heat exposure. Time-series modeling of temperature surfaces and mortality data shows that the deadliest heat events are often those characterized by multi-day persistence rather than extreme single-day peaks (Anderson & Bell, 2011; Vicedo-Cabrera et al., 2018). Incorporating temporal dimensions into GIS, through techniques such as stacked rasters or event-based composites, allows analysts to distinguish chronic hotspots from episodic extremes and to assess how risk evolves under changing climate conditions.

Recent advances in machine learning and data assimilation are expanding the scope of urban heat modeling, particularly in data-poor environments. Supervised classification algorithms have been used to identify heat-prone urban forms from high-resolution imagery, while downscaling techniques combine coarse climate model outputs with local land-cover data to estimate neighborhood-scale heat exposure (Perkins-Kirkpatrick & Lewis, 2020). Importantly for the scope of this book, these approaches can extend heat analysis to cities lacking dense meteorological networks, although they require careful validation and clear communication of uncertainty.

Ultimately, geospatial modeling of heat is not an exercise in prediction alone. Its value lies in making visible the spatial consequences of accumulated planning decisions, overlaid on infrastructure investment and persisting social inequalities. By linking surface temperature patterns to vegetation loss, housing quality, labor conditions, and access to services, heat maps become arguments; that is, evidence that extreme heat is neither natural nor evenly distributed but produced through identifiable and modifiable processes. The next section builds on this analytical foundation by examining how these models are applied in practice, using case studies and workflows that demonstrate how freely available data can support actionable heat mitigation strategies.

7.3.1 Urban Heat Island (UHI) Mapping Workflow

7.3.1.1 New Delhi Case Study—Raster + District Analysis

7.3.1.1.1 Overview

This workflow identifies and compares urban heat island conditions across legislative districts using freely available raster datasets. The approach combines LST, vegetation cover (NDVI), built-up land cover, and population density to highlight districts where heat exposure and human vulnerability intersect.

7.3.1.1.2 Step 1. Define the Study Area and Administrative Units

1. Select the city of interest and the administrative unit for comparison (e.g., legislative districts).
2. Acquire administrative boundary data from an open-source portal such as the Humanitarian Data Exchange (HDX).
3. Use these boundaries consistently to clip all raster datasets and to aggregate results.

7.3.1.1.3 Step 2. Acquire Core Raster Datasets

The following datasets are required:

1. Land Surface Temperature (LST)
   - Source: *Landsat 8* (30 m resolution)
   - Acquisition method: Google Earth Engine (GEE)
   - Temporal selection: Imagery from May 27, 2024, immediately prior to the 2024 Indian heat wave
   - Processing: Clip imagery to the New Delhi administrative boundary
2. Vegetation Cover (NDVI)
   - Source: *Landsat 8* Red and Near-Infrared (NIR) bands (30 m)
   - Processing: NDVI calculated in GEE using standard band math
   - Output clipped to administrative boundary
3. Population Data
   - Source: *WorldPop*
   - Resolution: 100 m population count raster (2020)
   - Processing: Downloaded and clipped in QGIS
4. Built-up Land Cover
   - Source: ESRI global Land Use/Land Cover dataset (10 m, 2024 composite)
   - Processing: Downloaded via GEE
   - Reclassified into two classes: built and not built

7.3.1.1.4 Step 3. Preprocess Raster Layers

1. Clip all raster datasets to the New Delhi boundary (performed in GEE for large imagery; QGIS for population raster).
2. Reproject rasters to a common coordinate reference system using Warp (Reproject) in QGIS.
3. For land-use data, reclassify LULC categories into a binary built/not-built raster.
4. Visually inspect each raster to confirm spatial alignment and expected value ranges.

7.3.1.1.5 Step 4. Aggregate Raster Values by District (Zonal Statistics)

1. Load legislative district boundaries and all raster layers into QGIS.
2. Use the Zonal Statistics tool to calculate district-level statistics for:
    - Mean LST
    - Mean NDVI
    - Mean built-up percentage
    - Population count
3. Calculate population density for each district using the field calculator:

$$\text{Population Density} = \frac{\text{Population Count}}{\text{District Area}}$$

7.3.1.1.6 Step 5. Produce Comparative District-Level Maps

1. Create four district-level choropleth maps:
    - Average Land Surface Temperature
    - Population Density
    - Built-up Percentage
    - Average NDVI
2. Symbolize each map using Natural Breaks (Jenks) classification.
3. Add a muted basemap to provide geographic context without competing visually.
4. Use callouts or arrows to highlight:
    - Hottest districts
    - Densely populated districts
    - Districts with low NDVI and high greening potential

7.3.1.1.7 Step 6. Interpret Patterns and Identify Priority Areas

1. Compare maps to identify districts where:
    - High LST overlaps with high population density

- High built-up percentages coincide with low vegetation

2. These districts represent priority zones for heat mitigation interventions such as:
   - Tree planting
   - Shading infrastructure
   - Cooling centers or water-based cooling strategies

7.3.1.1.8 Step 7. Extensions and Advanced Analyses (Optional)

This workflow represents a screening-level analysis. Potential extensions include:

- Calculation of the Urban Fabric Thermal Vulnerability Index (UFTVI)
- Regression analysis to quantify relationships between LST and other variables
- Time-series analysis across multiple heat events
- Integration with health or mortality data where available

Figure 7.2 depicts the variables NDVI, land surface temperatures, land use/land cover, and population density for New Delhi. Figure 7.3 depicts NDVI, land surface temperatures, land use/land cover, and built-up area by electoral district for New Delhi

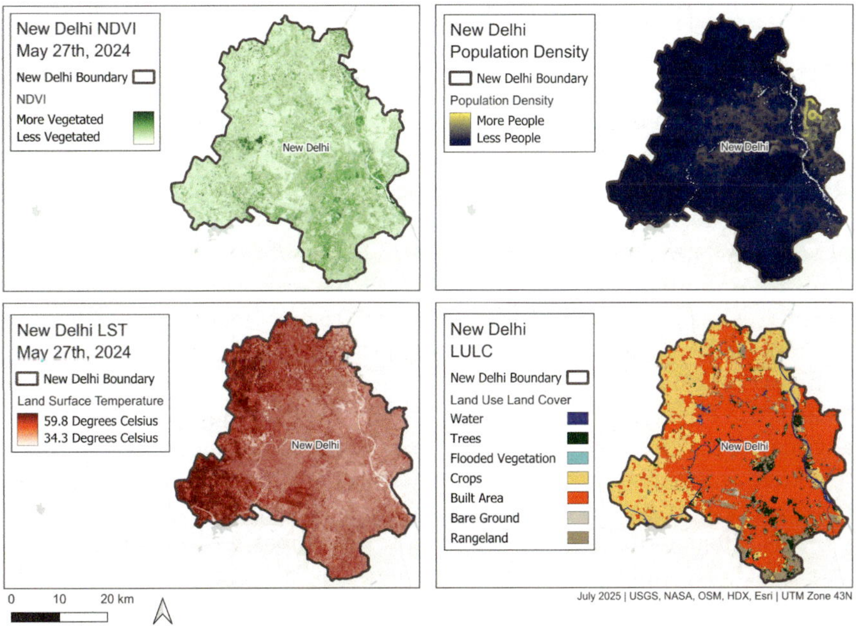

FIGURE 7.2
New Delhi NDVI LST PopDensity LULC.

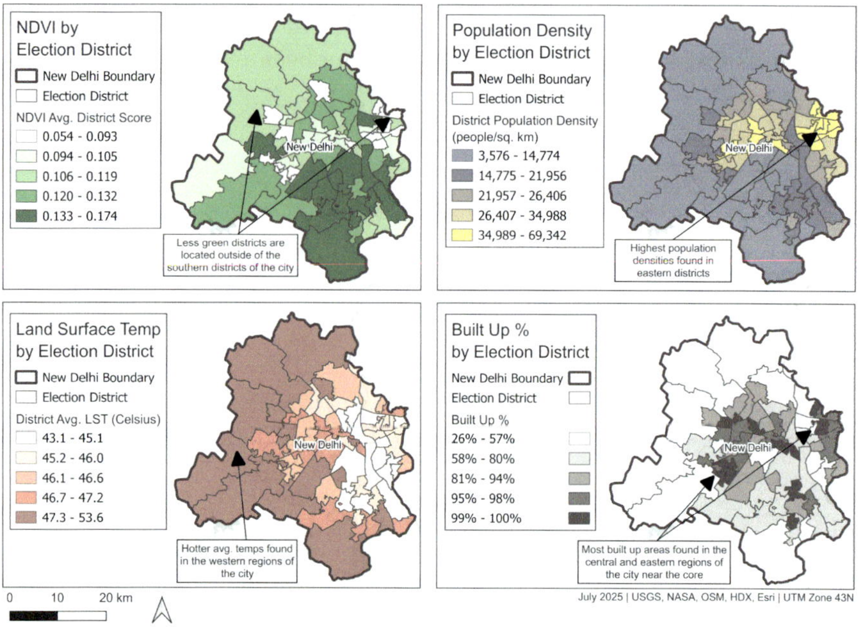

FIGURE 7.3
New Delhi NDVI LST PopDensity BuiltUp% by electoral district. (Cartography by John Lesko.)

7.4 Getting to Action: Mapping Heat Risk into Urban Intervention

Mapping urban heat is not an end in itself. The analytic value of heat mapping lies in its ability to inform concrete interventions—design, policy, and governance decisions that reduce exposure, protect vulnerable populations, and increase urban thermal resilience. Moving from spatial analysis to action requires more than identifying hotspots. It requires translating patterns of thermal exposure and vulnerability into decisions about where resources are deployed, which neighborhoods are prioritized, and how urban form is reshaped over time.

Geospatial analysis plays a central role in this translation because it links physical heat exposure to population distribution, housing conditions, infrastructure access, and institutional capacity. When these layers are analyzed together, heat maps become decision-support tools rather than descriptive artifacts. They allow cities to identify priority zones for intervention, evaluate trade-offs among competing strategies, and target limited resources where they are likely to yield the greatest public health benefit.

The sections that follow examine how heat mapping informs urban intervention across multiple scales, from neighborhood-level greening and cooling infrastructure to building design, regulatory frameworks, and public health preparedness. Throughout, the emphasis is on spatially targeted action: interventions that respond not to average citywide conditions, but to the highly uneven geography of heat risk within cities.

7.4.1 From Heat Surfaces to Priority Zones

Urban heat maps—whether derived from land surface temperature, air temperature interpolation, or composite vulnerability indices—reveal stark spatial heterogeneity. However, temperature alone is insufficient as a basis for intervention. Effective prioritization emerges when thermal exposure is analyzed in conjunction with population density, housing conditions, access to services, and adaptive capacity. Districts that are simultaneously hot, densely populated, and poorly vegetated represent fundamentally different planning challenges than hot but sparsely populated industrial zones (Harlan et al., 2013; Shandas et al., 2019).

Geospatial overlays allow planners to identify these compound risk zones explicitly. For example, combining LST rasters with population density surfaces and vegetation indices can reveal neighborhoods where modest temperature reductions would yield disproportionately large public health benefits. This logic mirrors approaches used in floodplain management and air pollution control, where exposure and vulnerability—not hazard magnitude alone—drive prioritization (IPCC, 2022).

7.4.2 Urban Greening as a Spatial Intervention

Urban greening remains one of the most robust and evidence-backed strategies for heat mitigation. Tree canopy, green corridors, and permeable surfaces reduce surface and air temperatures through shading and evapotranspiration, often lowering local temperatures by several degrees Celsius (Bowler et al., 2010; Ziter et al., 2019). Crucially, these benefits are highly localized, making spatial targeting essential.

GIS enables planners to identify where greening will be most effective by:

1. Locating districts with low NDVI and high thermal exposure
2. Identifying publicly owned or underutilized land suitable for intervention
3. Modeling canopy expansion scenarios and their potential cooling effects

Equity considerations are central. In many cities, historic disinvestment has left low-income neighborhoods with minimal tree cover and limited access to parks. Heat and UHI mapping exposes these inequities and provides empirical support for redistributive greening strategies that prioritize thermal justice rather than aesthetic enhancement (Hoffman et al., 2020).

7.4.3 Cooling Infrastructure and Access Mapping

Where greening alone is insufficient (or not feasible in the short term), cities increasingly rely on cooling infrastructure such as cooling centers, misting stations, shaded transit stops, public water fountains, and public water access. Geospatial analysis plays a critical role in determining whether these resources are accessible to those most at risk.

Network analysis can be used to assess walking or transit access to cooling centers under extreme heat conditions, incorporating travel time, slope, and barriers such as major roadways. Overlaying these accessibility surfaces with demographic data reveals gaps where vulnerable populations lack realistic access to cooling during heat emergencies (Uejio et al., 2011; McGeehin & Mirabelli, 2001).

Importantly, access is not purely spatial. Social factors depend on awareness based in trust and cultural norms, and they also shape whether cooling infrastructure is used. While GIS cannot capture these dynamics directly, spatial analysis can highlight where outreach and community-based adaptation efforts should be concentrated.

7.4.4 Housing, Building Form, and Regulatory Levers

Heat mapping also informs interventions at the scale of buildings and neighborhoods. Roof materials, insulation quality, ventilation, and building orientation strongly influence indoor thermal conditions. Cool reflective roofs and other building materials, along with passive ventilation strategies, have demonstrated measurable reductions in indoor temperatures, particularly in low-rise residential structures common in informal and peri-urban areas (Santamouris, 2014; Tuholske et al., 2021).

Spatial datasets on building footprints, roof type, and construction materials are more readily available now and can be integrated with heat exposure maps to identify housing stock most in need of retrofitting. In contexts where detailed building data are lacking, proxies such as settlement type, density, and roof reflectance derived from satellite imagery can still support targeted interventions.

Regulatory tools matter as much as technical ones. The development of zoning and building codes, as well as land-use regulations, shape urban thermal outcomes over decades. Heat maps provide empirical justification for revising these frameworks and using them proactively to encourage tree preservation, or limit green-area/impervious surface ratios, or incentive reflective materials. These actions will be particularly effective in rapidly

7.4.5 Early Warning, Public Health, and Governance

Finally, heat mapping contributes to early warning systems and public health preparedness. When combined with weather forecasts, historical mortality data, and vulnerability indicators, spatial heat analyses can support government (or volunteer) actions including targeted alerts, resource deployment, and emergency planning. Several cities now use heat-health warning systems that incorporate spatial risk to tailor responses by neighborhood rather than issuing citywide advisories (Ebi et al., 2021).

Governance capacity ultimately determines whether these tools translate into reduced harm. Heat is a slow-onset disaster that rarely triggers the same political urgency as floods or storms. Maps that make heat visible—spatially, socially, and temporally—are therefore essential political instruments. In this sense, heat mapping is not simply a diagnostic exercise, but a means of making extreme heat legible as a governable urban risk—one that reflects past planning choices and can still be reshaped through deliberate, spatially informed action.

References

Adams, Q.H., Chan, E.M.G., Spangler, K.R., Weinberger, K.R., Lane, K.J., Errett, N.A, Hess, J.J., Sun, Y., Wellenius, G.A., & Nori-Sarma, A (2023). Examining the Optimal Placement of Cooling Centers to Serve Populations at High Risk of Extreme Heat Exposure in 81 US Cities. *Public Health Rep.* Nov-Dec; *138*(6):955-962. https://doi.org/10.1177/00333549221148174.

Anderson, G. B., & Bell, M. L. (2011). Heat waves in the United States: Mortality risk during heat waves and effect modification by heat wave characteristics in 43 U.S. communities. *Environmental Health Perspectives, 119*(2), 210–218. https://doi.org/10.1289/ehp.1002313

Arnfield, A. J. (2003). Two decades of urban climate research: A review of turbulence, exchanges of energy and water, and the urban heat island. *International Journal of Climatology, 23*(1), 1–26. https://doi.org/10.1002/joc.859

Basu, R., & Samet, J. M. (2002). Relation between elevated ambient temperature and mortality: A review of the epidemiologic evidence. *Epidemiologic Reviews, 24*(2), 190–202. https://doi.org/10.1093/epirev/mxf007

Bowler, D. E., Buyung-Ali, L., Knight, T. M., & Pullin, A. S. (2010). Urban greening to cool towns and cities: A systematic review of the empirical evidence. *Landscape and Urban Planning, 97*(3), 147–155. https://doi.org/10.1016/j.landurbplan.2010.05.006

Ebi, K. L., Capon, A., Berry, P., Broderick, C., de Dear, R., Havenith, G., Honda, Y., Kovats, R. S., Ma, W., Malik, A., Morris, N. B., Nybo, L., Seneviratne, S. I., Vanos, J., & Jay, O. (2021). Heat watch–warning systems in a changing climate. *International Journal of Environmental Research and Public Health, 18*(8), 4212. https://doi.org/10.3390/ijerph18084212

Gasparrini, A., Guo, Y., Hashizume, M., Lavigne, E., Zanobetti, A., Schwartz, J., Tobias, A., Tong, S., Rocklöv, J., Forsberg, B., Leone, M., De Sario, M., Bell, M. L., Guo, Y.-L. L., Wu, C.-F., Kan, H., Yi, S.-M., de Sousa Zanotti Stagliorio Coelho, M., Saldiva, P. H. N., Honda, Y., Kim, H., & Armstrong, B. (2015). Mortality risk attributable to high and low ambient temperature: A multicountry observational study. *The Lancet, 386*(9991), 369–375. https://doi.org/10.1016/S0140-6736(14)62114-0

Harlan, S. L., Declet-Barreto, J. H., Stefanov, W. L., & Petitti, D. B. (2013). Neighborhood effects on heat deaths: Social and environmental predictors of vulnerability in Maricopa County, Arizona. *Environmental Health Perspectives, 121*(2), 197–204. https://doi.org/10.1289/ehp.1104625

Hoffman, J. S., Shandas, V., & Pendleton, N. (2020). The effects of historical housing policies on resident exposure to intra-urban heat: A study of 108 U.S. urban areas. *Climate, 8*(1), 12. https://doi.org/10.3390/cli8010012

Howard, L. (1833). *The climate of London, deduced from meteorological observations*. Harvey and Darton.

India Meteorological Department. (2024). *Press releases and observational summaries: May–June 2024 heat wave*. Government of India.

Intergovernmental Panel on Climate Change (IPCC). (2022). *Climate change 2022: Impacts, adaptation and vulnerability* (Sixth Assessment Report, Working Group II). Cambridge University Press.

Kenny, G. P., Yardley, J., Brown, C., Sigal, R. J., & Jay, O. (2017). Heat stress in older individuals and patients with common chronic diseases. *Canadian Medical Association Journal, 182*(10), 1053–1060. https://doi.org/10.1503/cmaj.081050

Kjellstrom, T., Freyberg, C., Lemke, B., Otto, M., & Briggs, D. (2018). Estimating population heat exposure and impacts on working people in conjunction with climate change. *International Journal of Biometeorology, 62*(3), 291–306. https://doi.org/10.1007/s00484-017-1407-0

McGeehin, M. A., & Mirabelli, M. (2001). The potential impacts of climate variability and change on temperature-related morbidity and mortality in the United States. Environmental Health Perspectives, *109*(Suppl 2), 185–189. https://doi.org/10.1289/ehp.109-1240665

Oke, T. R. (1982). The energetic basis of the urban heat island. *Quarterly Journal of the Royal Meteorological Society, 108*(455), 1–24. https://doi.org/10.1002/qj.49710845502

Oke, T. R. (1987). *Boundary layer climates* (2nd ed.). Routledge.

Parsons, K. (2014). *Human thermal environments: The effects of hot, moderate, and cold environments on human health, comfort, and performance* (3rd ed.). CRC Press.

Perkins-Kirkpatrick, S. E., & Lewis, S. C. (2020). Increasing trends in regional heatwaves. *Nature Communications, 11*, 3357. https://doi.org/10.1038/s41467-020-16970-7

Raymond, C., Matthews, T., & Horton, R. M. (2020). The emergence of heat and humidity too severe for human tolerance. *Science Advances, 6*(19), eaaw1838. https://doi.org/10.1126/sciadv.aaw1838

Reid, C. E., O'Neill, M. S., Gronlund, C. J., Brines, S. J., Brown, D. G., Diez-Roux, A. V., & Schwartz, J. (2009). Mapping community determinants of heat vulnerability. Environmental Health Perspectives, *117*(11), 1730–1736. https://doi.org/10.1289/ehp.0900683

Robine, J. M., et al. (2008). Death toll exceeded 70,000 in Europe during the summer of 2003. *Comptes Rendus Biologies, 331*(2), 171–178. https://doi.org/10.1016/j.crvi.2007.12.001

Roth, M. (2007). Review of urban climate research in (sub)tropical regions. *International Journal of Climatology, 27*(14), 1859–1873. https://doi.org/10.1002/joc.1591

Santamouris, M. (2014). Cooling the cities—A review of reflective and green roof mitigation technologies. *Solar Energy, 103*, 682–703. https://doi.org/10.1016/j.solener.2012.07.003

Sera, F., Armstrong, B., Tobias, A., Vicedo-Cabrera, A. M., Åström, C., Bell, M. L., Chen, B.-Y., de Sousa Zanotti Stagliorio Coelho, M., Correa, P. M., Dang, T. N., Diaz, M. H., Dung, D. V., Forsberg, B., Goodman, P. G., Guo, Y.-L. L., Guo, Y., Hashizume, M., Honda, Y., Hurtado-Diaz, M., Kim, H., Lavigne, É., Michelozzi, P., Milojevic, A., Overcenco, A., Ortega, N. V., Pascal, M., Ragettli, M. S., Ryti, N. R. I., Scovronick, N., Tong, S., Williams, G., Zanobetti, A., Gasparrini, A., & Schwartz, J. (2019). How urban characteristics affect vulnerability to heat and cold: A multi-country analysis. *International Journal of Epidemiology, 48*(4), 1101–1112. https://doi.org/10.1093/ije/dyz008

Sherwood, S. C., & Huber, M. (2010). An adaptability limit to climate change due to heat stress. *Proceedings of the National Academy of Sciences, 107*(21), 9552–9555. https://doi.org/10.1073/pnas.0913352107

Shandas, V., Voelkel, J., Williams, J., & Hoffman, J. (2019). Integrating satellite and ground measurements for predicting locations of extreme urban heat. *Climate, 7*(1), 5. https://doi.org/10.3390/cli7010005

Stewart, I. D., & Oke, T. R. (2012). Local climate zones for urban temperature studies. *Bulletin of the American Meteorological Society, 93*(12), 1879–1900. https://doi.org/10.1175/BAMS-D-11-00019.1

Tuholske, C., Caylor, K., Funk, C., Verdin, A., Sweeney, S., Grace, K., Peterson, P., & Evans, T. (2021). Global urban population exposure to extreme heat. *Proceedings of the National Academy of Sciences, 118*(41), e2024792118. https://doi.org/10.1073/pnas.2024792118

Uejio, C. K., Wilhelmi, O. V., Golden, J. S., Mills, D. M., Gulino, S. P., & Samenow, J. P. (2011). Intra-urban societal vulnerability to extreme heat. *Environmental Research, 111*(4), 507–517. https://doi.org/10.1016/j.envres.2011.02.005

Vandentorren, S., Suzan, F., Medina, S., Pascal, M., Maulpoix, A., Cohen, J.-C., & Ledrans, M. (2006). Mortality in 13 French cities during the August 2003 heat wave. American Journal of Public Health, 96(9), 1518–1520. https://doi.org/10.2105/ajph.94.9.1518

Vicedo-Cabrera, A. M., Sera, F., Gasparrini, A., Handschin, S., Guo, Y., Tong, S., Overcenco, A., de Sousa Zanotti Stagliorio Coelho, M., Saldiva, P. H. N., Lavigne, É., Correa, P. M., Ortega, N. V., Kan, H., Osorio, S., Jaakkola, J. J. K., Ryti, N. R. I., Goodman, P. G., Zeka, A., Michelozzi, P., Scovronick, N., Hashizume, M., Honda, Y., Kim, H., Tobias, A., Iñiguez, C., Forsberg, B., Åström, C., Ragettli, M. S., Guo, Y.-L. L., Chen, B.-Y., Zanobetti, A., Schwartz, J., Armstrong, B., & Gasparrini, A. (2018). Temperature-related mortality impacts under and beyond Paris Agreement climate change scenarios. *Climatic Change, 150*, 391–402. https://doi.org/10.1007/s10584-018-2274-3

Voogt, J. A., & Oke, T. R. (2003). Thermal remote sensing of urban climates. *Remote Sensing of Environment, 86*(3), 370–384. https://doi.org/10.1016/S0034-4257(03)00079-8

Weng, Q. (2009). Thermal infrared remote sensing for urban climate and environmental studies: Methods, applications, and trends. *ISPRS Journal of Photogrammetry and Remote Sensing, 64*(4), 335–344. https://doi.org/10.1016/j.isprsjprs.2009.03.007

Wolf, T., McGregor, G., & Analitis, A. (2015). Performance assessment of a heat wave vulnerability index for Greater London, United Kingdom. Weather, Climate, and Society, *7*(1), 32–46. https://doi.org/10.1175/WCAS-D-13-00014.1

World Health Organization. (2023). *Climate change, heat and health*. WHO.

Zhou, D., Xiao, J., Bonafoni, S., Berger, C., Deilami, K., Zhou, Y., Frolking, S., Yao, R., Qiao, Z., & Sobrino, J. A. (2017). Satellite remote sensing of surface urban heat islands: Progress, challenges, and perspectives. Remote Sensing, *9*(7), 675. https://doi.org/10.3390/rs9070675

Ziter, C. D., Pedersen, E. J., Kucharik, C. J., & Turner, M. G. (2019). Scale-dependent interactions between tree canopy cover and impervious surfaces reduce daytime urban heat during summer. *Proceedings of the National Academy of Sciences, 116*(15), 7575–7580. https://doi.org/10.1073/pnas.1817561116

8

Mapping Cyclones, Wind, and Urban Risk

Chapter Introduction

Cyclones, which are also known regionally as hurricanes or typhoons, are among the most destructive environmental threats affecting cities worldwide. They operate as compound hazards, combining extreme winds, heavy rainfall, and storm surges in ways that overwhelm infrastructure systems, disrupt livelihoods, and expose long-standing inequalities in urban form and governance (Emanuel, 2005; IPCC, 2022). A growing body of research indicates that while the total number of cyclones may not be increasing globally, their intensity and destructive potential are rising as oceans warm and atmospheric circulation patterns shift (Knutson et al., 2020). In urban settings, these changes translate into higher wind loads on buildings and greater levels of infrastructure failure.

For cities, wind is rarely the sole source of danger. Cyclonic storms damage buildings, uproot trees, sever power and communications networks, contaminate water supplies, and trigger cascading hazards such as flooding, landslides, and disease outbreaks. These impacts are profoundly spatial. Exposure varies with factors such as elevation, proximity to the coast, surface roughness, building quality, land use, and access to safe shelter. Informal settlements and peripheral neighborhoods often experience the greatest damage, not only because of their location but also because of cumulative vulnerabilities produced by planning decisions and uneven governance (Pelling, 2011).

This chapter examines cyclones as an urban wind threat through a geospatial lens. It shows how wind exposure and storm impacts can be modeled using spatial data, and how these models can be used to identify zones of heightened risk within cities. This chapter opens with the experience of a tropical cyclone in the South Pacific, which serves as a stark illustration of how a single wind event can reshape an urban future. It then unpacks the physical science of cyclones, introduces geospatial approaches to modeling wind and storm exposure, and concludes by examining how spatial analysis can inform preparedness, mitigation, and recovery in cities facing increasingly volatile wind regimes.

DOI: 10.1201/9781032636948-8

8.1 Setting the Scene: When Wind Becomes Catastrophe

In February 2016, Tropical Cyclone Winston made landfall in Fiji with sustained wind speeds exceeding 230 km h^{-1} (145 mph), making it the strongest cyclone ever recorded in the Southern Hemisphere at the time (World Meteorological Organization, 2017). The storm's winds rather than flooding alone were responsible for much of the devastation. Entire villages were flattened as lightweight housing structures failed under sustained gusts and extreme pressure differentials. Electrical transmission towers collapsed, communications infrastructure was destroyed, and road networks were rendered impassable by wind-thrown debris. In the immediate aftermath, approximately 62 percent of Fiji's population, which was more than 540,000 people, were affected, with tens of thousands displaced and essential services interrupted nationwide (Government of Fiji, 2016).

Unlike floodwaters, which often recede within hours or days, wind damage is instantaneous and structural. During Cyclone Winston, wind forces stripped roofs from homes, shattered windows, uprooted mature trees, and caused cascading failures in energy and telecommunications systems. In many coastal and island communities, housing stock consisted of timber frames, corrugated metal roofing, and unreinforced connections between roofs, walls, and foundations. Such construction materials are particularly vulnerable to uplift and lateral wind loads (Paulay & Priestley, 1992; FEMA, 2012). These failures were not the result of poor craftsmanship alone, but of systemic mismatches between the existing building practices and the intensifying wind regimes of a warming climate.

Wind hazards of this magnitude are not random anomalies. Tropical cyclones derive their energy from warm ocean waters and organized atmospheric circulation, converting latent heat into mechanical energy that manifests as extreme winds. Sea surface temperatures exceeding approximately 26.5°C (79.7°F) are a prerequisite for cyclone formation, while low vertical wind shear allows storms to intensify and maintain structural coherence (Emanuel, 2005). During the 2015–2016 El Niño event, anomalously warm waters in the South Pacific created conditions highly favorable for rapid cyclone intensification, contributing to Winston's exceptional strength (Australian Bureau of Meteorology, 2016). Climate change is now shifting these thresholds, expanding the spatial and temporal envelope in which destructive wind events can occur.

What makes wind a particularly dangerous urban threat is its selective destructiveness. Within the same city or island, damage patterns often vary dramatically over short distances. Exposure depends on factors such as building orientation, elevation, surface roughness, and proximity to open water, which allows wind to accelerate over long fetch distances. Informal settlements and peripheral neighborhoods, which are almost never planned but are constructed incrementally without engineering oversight, tend to experience disproportionate damage, sometimes while nearby reinforced commercial or governmental structures remain intact. This uneven geography of destruction translates directly into unequal recovery trajectories, as communities with fewer resources face longer rebuilding times and greater livelihood disruption (Pelling, 2011).

Wind also operates synergistically with other hazards. Even where storm surge and rainfall are limited, wind damage alone can precipitate humanitarian crises by disabling power grids, contaminating water supplies, and severing transportation corridors. Hospitals and emergency shelters may be structurally sound yet rendered nonfunctional by the loss of electricity or communications. In Fiji, nearly 80 percent of the population lost power, including entire islands that remained disconnected for weeks (Government of Fiji, 2016). Such cascading failures underscore the importance of treating wind not as a secondary effect of storms, but as a primary driver of urban risk.

Although cyclones are most associated with tropical regions, extreme wind hazards are not geographically confined. Mid-latitude cities increasingly experience severe convective wind events, including derechos, downbursts, and tornado outbreaks, which produce damage patterns comparable to tropical cyclones (Trapp et al., 2016). In these contexts, wind acts on dense urban infrastructure that was never designed for sustained high-velocity gusts, exposing vulnerabilities in aging buildings, overhead utilities, and other elements of the urban landscape. As climate change alters atmospheric instability and moisture availability, the frequency and intensity of such wind events are projected to increase in many regions (IPCC, 2022).

This chapter approaches wind as a spatially structured environmental threat that can be analyzed, visualized, and mitigated using geospatial tools. Rather than treating wind damage as an inevitable consequence of extreme weather, it emphasizes the role of exposure, compounded by built form, and their roles in shaping outcomes. By integrating historical storm tracks, population distributions, land cover, and wind, GIS enables analysts to move beyond post-disaster mapping toward proactive risk identification and resilience planning. The sections that follow unpack the science of wind hazards, examine their uneven impacts on cities, and demonstrate practical workflows for mapping wind risk using open data.

8.2 The Science of Cyclones and Wind

Cyclones are large-scale atmospheric systems characterized by closed low-pressure circulation, organized convection, and sustained high winds. In tropical regions, these systems derive their energy primarily from the release of latent heat associated with deep convection over warm ocean surfaces.

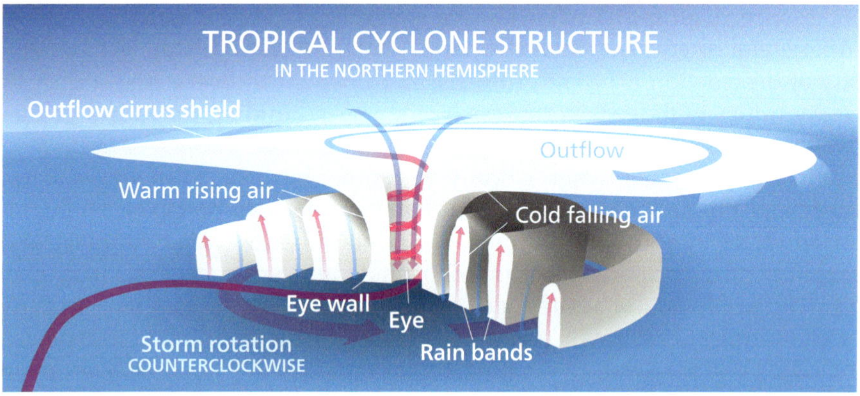

FIGURE 8.1
Structure of a cyclone: Eye and eyewall. Public Domain. (Tropical Cyclone Structure (in en). *National Oceanic and Atmospheric Administration* (2022-11-10). Structure and Motion (in en). *Environment Canada* (2013-07-23).)

Sea surface temperatures above approximately 26°C–27°C (80°F) provide the thermal conditions necessary for cyclone genesis, while weak vertical wind shear allows storms to intensify and maintain vertical coherence (Gray, 1979; Emanuel, 1986). Once established, cyclones function as heat engines, converting oceanic thermal energy into kinetic energy expressed as wind. Figure 8.1 illustrates the structure of a tropical cyclone, showing the eye, the eyewall, rain bands, outflow, and storm rotation.

Wind within a cyclone is governed by the interaction of pressure gradients, the Coriolis effect, and surface friction. Near the Earth's surface, air flows inward toward the low-pressure center, accelerating as pressure gradients steepen. Friction reduces wind speed close to the ground, causing air to spiral inward rather than circulate symmetrically around the storm center. Above the boundary layer, winds approach gradient balance, producing the strongest sustained velocities near the radius of maximum winds surrounding the storm's eye (Holton & Hakim, 2013). These dynamics generate wind fields that are spatially uneven and highly sensitive to surface conditions after landfall.

Wind damage in urban areas is shaped by both wind speed and duration, as well as by short-lived gusts. While sustained winds determine overall storm severity, gusts frequently cause the most acute structural damage by exceeding design thresholds for buildings, trees, and infrastructure (Vickery et al., 2006). Gustiness arises from turbulence within the cyclone boundary layer and from interactions between atmospheric flow and surface roughness elements such as buildings, vegetation, and terrain. As cyclones move inland, increased surface friction reduces average wind speed but enhances turbulence, which can locally intensify gust impacts in dense urban districts.

Urban form plays a critical role in shaping wind exposure at neighborhood scales. Building height, spacing, orientation, and construction quality influence how wind energy is redistributed once it enters the urban canopy layer. High-rise structures can channel winds downward through downdrafts and corner accelerations, intensifying near-surface wind speeds along streets and open spaces. In contrast, tightly packed low-rise neighborhoods often experience widespread structural failure due to cumulative stress on weaker buildings rather than isolated zones of extreme wind (Tamura et al., 2015). These effects mean that cyclone wind risk cannot be inferred from regional wind speed alone. It must be understood as a spatially heterogeneous phenomenon shaped by urban morphology.

Topography further modifies cyclone wind fields following landfall. Coastal plains, escarpments, and mountain ranges can accelerate or decelerate winds depending on slope orientation and storm trajectory. Orographic lifting enhances precipitation on windward slopes, while downslope winds may intensify gusts on leeward sides through compression and channeling effects (Smith, 1979). In cities located near complex terrain, such as Port-au-Prince, Hong Kong, or Caracas, these interactions produce sharp gradients in wind damage over short distances, complicating emergency response and recovery. Figure 8.2 depicts the movement of air masses over a series of mountain ranges.

Although cyclones are often discussed primarily in terms of wind categories, wind rarely acts in isolation. High winds interact with rainfall, storm surge, and pre-existing environmental conditions to produce compound impacts. Wind-driven rain infiltrates buildings, accelerates material degradation, and compromises electrical systems. Fallen trees and debris obstruct

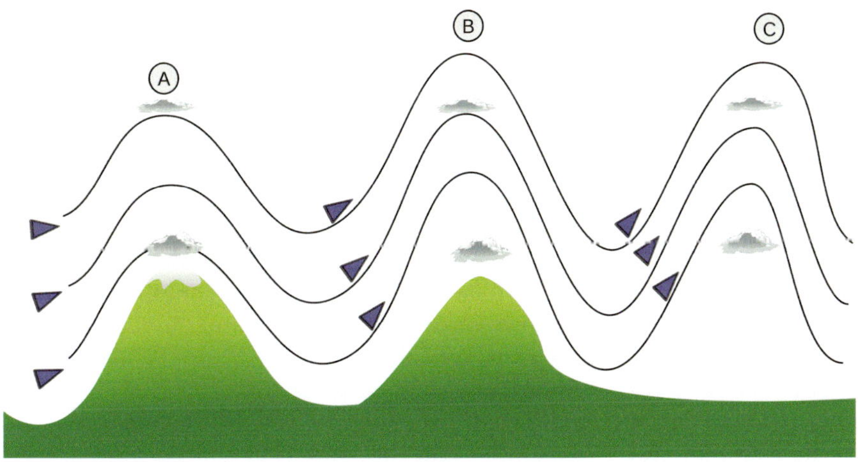

FIGURE 8.2
Wind–topography interaction. (Creative Commons. https://upload.wikimedia.org/wikipedia/commons/9/9a/Onde_orographique_r%C3%A9sonance.svg)

drainage networks, exacerbating flooding during and after storms. Power outages triggered by wind damage disable pumping stations, hospitals, and communications infrastructure, magnifying indirect losses long after peak winds subside (Cutter et al., 2008). From a geospatial perspective, these interactions underscore the need to treat wind as part of a coupled hazard system rather than as a standalone variable.

Climate change is altering cyclone wind characteristics in ways that are increasingly consequential for cities. While global trends in cyclone frequency remain uncertain, there is strong evidence that the proportion of high-intensity storms is increasing, along with peak wind speeds and storm-related rainfall (Knutson et al., 2020). Slower storm translation speeds observed in several regions further extend the duration of wind exposure over urban areas, increasing cumulative stress on buildings and infrastructure (Kossin, 2018). Many urban building standards, however, remain anchored in historical wind climatologies that no longer reflect emerging risk.

Understanding the physical science of cyclone winds is therefore foundational for spatial threat mapping. Wind risk emerges from the interaction of atmospheric dynamics, terrain, urban form, and infrastructure vulnerability. Geospatial analysis enables these components to be integrated, revealing where extreme winds are likely to concentrate, which neighborhoods are least able to withstand them, and how secondary impacts may cascade through urban systems. The next section translates this physical understanding into geospatial representations of wind exposure and damage potential.

8.3 Modeling Wind Risk in Geospatial Terms

Modeling cyclone wind risk in geospatial terms requires translating dynamic atmospheric processes into spatial representations that can be linked to urban form, infrastructure, and population exposure. Unlike hazards with fixed footprints, wind is inherently variable across space and time. Its intensity, direction, and duration change rapidly as storms move, interact with terrain, and encounter built environments. As a result, geospatial analysis of wind focuses less on predicting exact wind speeds at specific locations and more on identifying patterns of relative exposure, zones of amplification, and areas where wind interacts with social and infrastructural vulnerability.

At regional scales, wind hazard modeling typically begins with historical storm tracks and reanalysis datasets that describe storm position, intensity, size, and translation speed. Best-track datasets produced by agencies such as the National Hurricane Center and the Joint Typhoon Warning Center provide standardized records that can be interpolated to generate probabilistic wind surfaces (Landsea & Franklin, 2013). These data are often combined with parametric wind field models, such as the Holland model, which estimate spatial wind distributions based on storm central pressure, radius of maximum winds, and post-landfall decay (Holland, 1980; Vickery et al., 2006). While simplified, these approaches offer a tractable means of mapping wind exposure over large urban regions where direct observations are sparse.

Downscaling these regional wind fields to the urban scale introduces additional complexity. Terrain plays a central role in modifying wind speed and direction through orographic acceleration, channeling, and sheltering effects. Digital Elevation Models (DEMs) are therefore a foundational input for spatial wind analysis, enabling the identification of ridgelines, valleys, and escarpments that influence post-landfall wind behavior (Miller et al., 2013). In coastal cities adjacent to complex terrain, even modest elevation differences can produce sharp gradients in wind damage, underscoring the importance of high-resolution topographic data.

Urban morphology further shapes wind exposure at neighborhood and block scales. Building height, density, orientation, and porosity influence how wind energy is redistributed once it enters the urban canopy layer. Although computational fluid dynamics (CFD) models can simulate airflow around individual structures with high precision, they are computationally intensive and rarely feasible for citywide analysis, particularly in resource-constrained settings (Blocken, 2014). As a result, most urban-scale wind assessments rely on proxy variables derived from geospatial data, such as building density, floor area ratio, land-use class, and surface roughness length, to infer relative exposure and damage potential (Oke, 1987; Grimmond & Oke, 1999).

Satellite remote sensing plays a complementary role by providing spatial indicators of wind vulnerability rather than wind itself. High-resolution optical imagery and building footprint datasets can be used to map roof materials, building age, and informal construction patterns—factors strongly correlated with wind damage during cyclones (Pelling, 2011; Taubenböck et al., 2013). Nighttime lights data and electrical infrastructure layers offer additional insight into cascading impacts, as wind-induced power outages often serve as proxies for broader system failure (Arsenault et al., 2020). When integrated within a GIS environment, these layers allow analysts to move beyond abstract wind speeds toward spatially explicit representations of who and what is at risk.

Exposure modeling must also account for population distribution and mobility. Census data, gridded population estimates, and settlement layers enable the overlay of wind hazard surfaces with residential density, helping to identify neighborhoods where high winds intersect with large populations or limited evacuation capacity (Balk et al., 2006). In many cities around the globe, informal settlements are located on coastal margins or steep hillsides and exhibit a particularly dangerous vulnerability that represents a convergence of wind exposure, weak building standards, and limited access to safe shelter. Mapping these intersections is critical for prioritizing preparedness efforts and emergency response.

Temporal considerations are equally important. Wind risk is shaped not only by peak intensity but also by duration of exposure and timing relative to other hazards. Slower-moving storms subject urban areas to prolonged wind stress, increasing cumulative damage to buildings and infrastructure (Kossin, 2018). Geospatial workflows increasingly incorporate time-enabled datasets—such as storm position by hour or day—to visualize how wind exposure evolves as a cyclone approaches, stalls, and dissipates. While such analyses are often exploratory rather than predictive, they provide valuable context for understanding compound impacts and recovery trajectories.

From a methodological standpoint, wind threat mapping is best understood as a scenario-based exercise rather than a deterministic forecast. Analysts assemble plausible wind fields based on historical or synthetic storms, overlay them with spatial indicators of vulnerability, and evaluate relative risk across urban space. This approach aligns with disaster risk reduction frameworks that emphasize preparedness, robustness, and adaptive capacity over precise prediction (UNDRR, 2015). Importantly, it allows cities to identify "no-regret" interventions—such as strengthening critical facilities, reinforcing housing in high-exposure zones, and protecting lifeline infrastructure—that reduce risk across a wide range of storm scenarios.

In the context of this book, modeling wind geospatially serves two purposes. First, it provides a concrete illustration of how atmospheric science can be translated into actionable spatial analysis using widely available data. Second, it highlights the limits of purely physical models when dislocated from their social and infrastructural contexts. Wind maps become meaningful only when they are interpreted alongside information about human-scale variables such as governance and existing levels of social inequality. The following section builds on this foundation by examining a specific urban case, demonstrating how geospatial wind analysis can reveal both patterns of exposure and opportunities for intervention.

8.3.1 Geospatial Workflow: Mapping Cyclone Exposure and Population Risk in Dominíca

This workflow demonstrates how widely available geospatial datasets can be combined to create a rapid, spatially explicit assessment of cyclone exposure and population risk at the national scale. While the example focuses on Dominíca, the approach is transferable to other small island states and coastal cities exposed to tropical cyclones.

8.3.1.1 Step 1. Define the Study Area and Administrative Units

The first step is to establish a clear spatial frame of reference for analysis.

1. Obtain national and sub-national administrative boundaries (country and parish level).

2. Suitable sources include:
    - Humanitarian Data Exchange (HDX, UN OCHA)
    - National or regional open data portals
    - OpenStreetMap (OSM)
3. In this example, Domínica country and parish boundaries were extracted from OpenStreetMap using the QuickOSM plugin in QGIS.

Administrative units serve two purposes:

1. constraining spatial queries and raster processing, and
2. providing a reporting framework for population and exposure metrics.

8.3.1.2 Step 2. Identify Relevant Cyclone Tracks

Cyclone exposure is assessed using historical storm track data.

1. Acquire historical tropical cyclone tracks from authoritative global datasets.
    - For this analysis, NOAA's global hurricane tracks (1842–2024) accessed via Esri's Living Atlas were used.
2. Storm track datasets typically represent each storm as a sequence of line segments, with attributes describing:
    - Storm intensity (e.g., tropical depression to Category 5)
    - Central pressure
    - Wind speed
    - Timestamp

Because national-scale situational awareness does not require all regional storms, the dataset is spatially filtered.

8.3.1.3 Step 3. Select Storms Relevant to the Study Area

To isolate storms relevant to Dominíca:

1. Use the Extract Within Distance tool to identify storm track segments within a specified buffer distance (e.g., 50 miles) of the Dominíca boundary.
2. This operation produces a partial subset of track segments rather than full storm trajectories.

3. To restore full storm paths:
   - Identify the unique storm IDs present in the subset.
   - Apply an attribute filter to the original storm track dataset using those storm IDs.
4. The result is a filtered storm track layer containing complete trajectories for storms that passed near Dominíca.

This approach preserves storm evolution while focusing analysis on geographically relevant events.

8.3.1.4 Step 4. Symbolize Cyclone Intensity

Storm tracks are symbolized by intensity class to communicate risk escalation along each trajectory.

1. Use NOAA-defined storm categories (Tropical Depression through Category 5).
2. Apply a graduated or categorical symbology that visually conveys intensification over warm Caribbean waters.
3. This visualization highlights:
   - Typical storm genesis regions
   - Intensification zones
   - Relative severity of storms affecting Dominíca

8.3.1.5 Step 5. Prepare Population Data

Population exposure is represented using both census aggregates and fine-scale settlement indicators.

1. Acquire population data appropriate for the scale of analysis:
   - National census data where available
   - Global gridded datasets (e.g., WorldPop) where census data are unavailable or outdated
2. For this example:
   - 2011 Dominíca census data at the parish level were joined to parish polygons.
3. To improve intra-parish spatial resolution:
   - Building footprints were extracted from OpenStreetMap using QuickOSM (building=yes).
4. The combined layer allows analysts to distinguish between:
   - Parishes with high total population
   - Areas of concentrated settlement within parishes

8.3.1.6 Step 6. Incorporate Wind Exposure Data

Wind exposure is mapped using modeled climatological wind datasets.

1. Acquire wind speed or wind power density data.
    - For this analysis, the World Bank Global Wind Atlas (250 m resolution) was used.
2. These raster datasets identify areas historically exposed to high wind energy.
3. No additional preprocessing is required beyond:
    - Reprojection (if necessary)
    - Appropriate symbology to emphasize high-exposure zones

Wind layers provide a spatial proxy for structural damage risk during cyclones.

8.3.1.7 Step 7. Incorporate Precipitation Data

To capture compound hazards, precipitation is included alongside wind.

1. Acquire precipitation data from a country-specific or regional data portal.
    - In this case, precipitation data were sourced from DomiNode, Dominíca's national data portal.
2. Precipitation layers contextualize:
    - Flood risk
    - Landslide susceptibility
    - Post-storm infrastructure and health impacts

8.3.1.8 Step 8. Produce a Composite Map Series

The final analytical step is cartographic integration.

1. Create four complementary maps:
    - Historical cyclone tracks
    - Population distribution
    - Wind exposure
    - Precipitation patterns
2. Combine these maps into a single multi-panel layout in QGIS to provide a unified situational overview.
3. Optional enhancements include:
    - Charts showing temporal trends in storm frequency and intensity

- Annotations highlighting particularly vulnerable regions or historical events (e.g., Hurricane Maria)

Figure 8.3 shows the historical record of tropical storm and hurricane tracks intersecting with the island of Dominíca in the Caribbean.

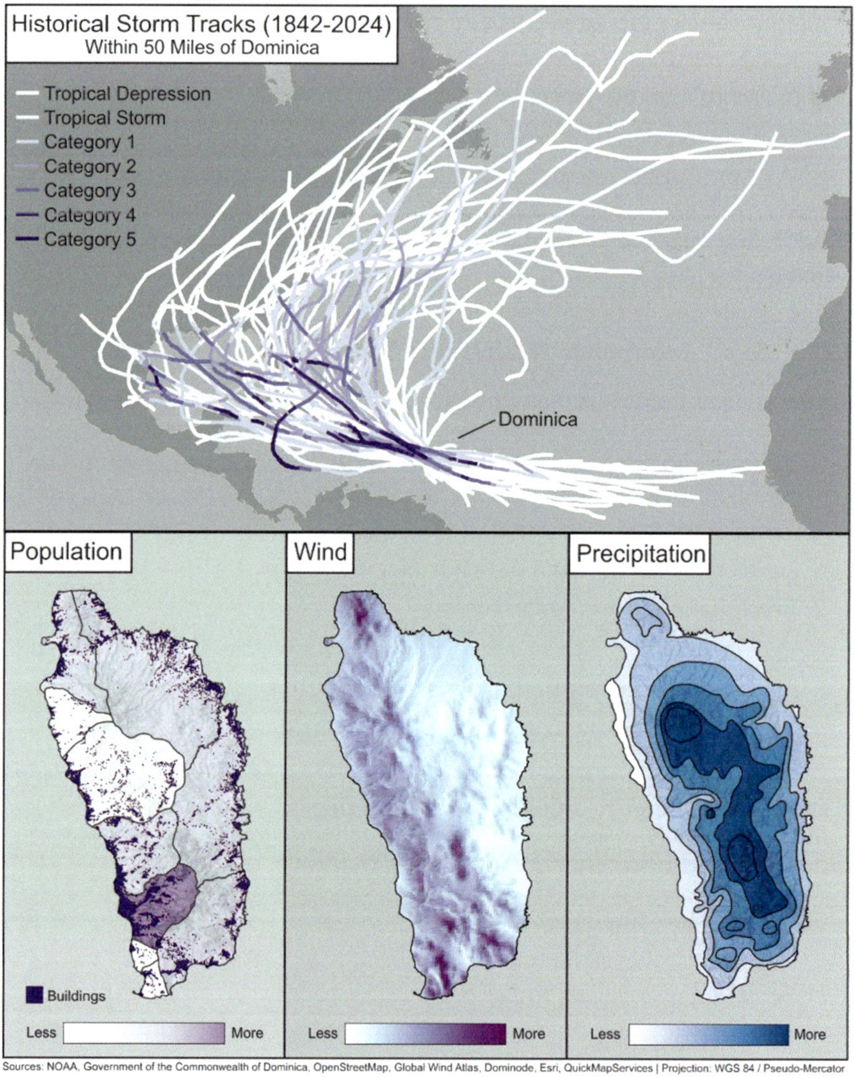

FIGURE 8.3
Dominíca historical storm track map. (Cartography by Maya O'Brien.)

8.3.1.9 Interpretation and Use

This map series supports rapid situational awareness by identifying where cyclone hazards intersect with population exposure and environmental conditions. While not predictive, the workflow enables decision-makers to:

- Prioritize preparedness and evacuation planning
- Identify parishes requiring targeted infrastructure reinforcement
- Contextualize recovery needs following major storms

The approach is intentionally modular, allowing analysts to substitute datasets based on local availability while preserving methodological consistency.

8.4 Getting to Action on Wind and Cyclone Risk

Extreme wind hazards demand forms of action that extend well beyond emergency response. Because cyclones are spatially extensive, recurrent, and increasingly intense, effective intervention depends on anticipatory planning, spatial targeting, and institutional coordination. Geospatial analysis plays a critical role in this transition from reaction to preparedness by making visible where wind risk concentrates, which populations and assets are most exposed, and how limited resources can be deployed strategically before a storm occurs.

8.4.1 From Hazard Mapping to Preparedness

One of the most consequential uses of wind-risk mapping is the identification of zones requiring advance planning rather than ad hoc response. Historical storm tracks, combined with modeled wind fields and topographic exposure, can delineate areas repeatedly affected by damaging winds. When overlaid with housing footprints and infrastructure networks, these layers reveal where emergency planning, early warning, and risk communication should bc prioritized before a cyclone forms.

Geospatial tools also support evacuation planning by integrating road networks, elevation, flood susceptibility, and distance-to-shelter metrics. These analyses are particularly important in island states and coastal cities, where a limited number of transport corridors may be vulnerable to wind damage, debris blockage, or secondary flooding. Mapping these constraints in advance allows authorities to test evacuation scenarios, identify bottlenecks, and issue clearer guidance under time pressure.

Critical infrastructure protection is another domain where spatial analysis enables action. Power substations, hospitals, water treatment facilities, telecommunications hubs, and emergency operations centers can be screened against modeled wind exposure and historical damage patterns. This information supports targeted reinforcement, redundancy planning, and restoration sequencing after storms. Without such spatial prioritization, recovery efforts risk reinforcing existing inequalities by restoring service first to politically or economically favored areas rather than to those with greatest need.

8.4.2 Translating Building Science into Spatial Policy

At the scale of individual structures, wind damage is governed by a small number of physical principles, including uplift forces, lateral pressure, load transfer, and connection failure. While building codes and engineering standards vary widely across countries, these principles are universal and can be translated into spatially informed policy guidance rather than prescriptive construction manuals.

Figure 8.4 depicts the pressure forces of high winds on a residential roof act on both the outside and inside of the structure.

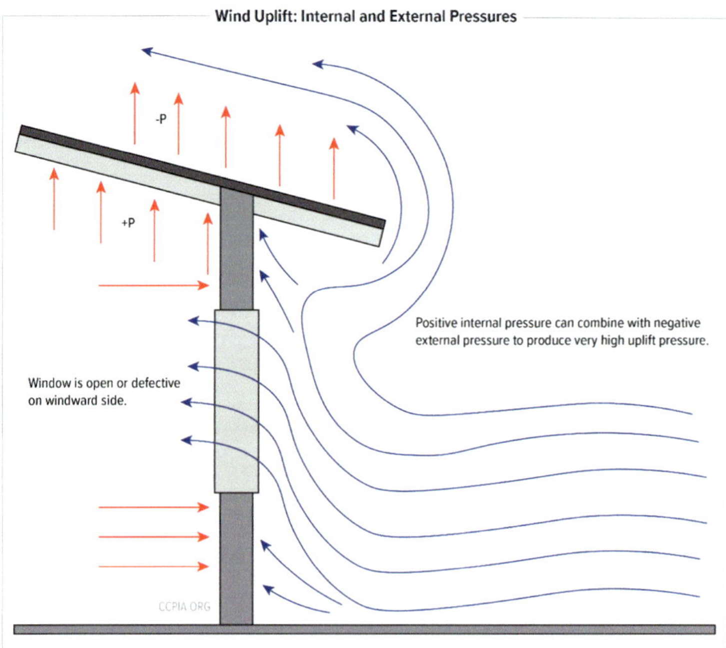

FIGURE 8.4
Roof failure diagram. (Wind pressure distribution around a ventilated roof space. Author: Bruce Taylor, Robert Gordon University. CC BY-SA https://commons.wikimedia.org/wiki/File:Wind_pressure_variation_ventilated_roof.png)

Roof failure is one of the most common entry points for catastrophic wind damage. Once a roof is compromised, internal pressurization sharply increases the likelihood of structural collapse. Geospatial inventories of building type, age, and roofing material—derived from cadastral records, remote sensing, or OpenStreetMap—allow cities to identify neighborhoods dominated by particularly vulnerable construction forms. These maps can inform targeted retrofit programs, insurance incentives, and post-disaster rebuilding guidelines.

Wall and frame performance is similarly dependent on continuity and load distribution. Structures with continuous sheathing and well-connected framing systems are better able to distribute wind forces and resist progressive failure. In many rapidly urbanizing cities, informal or semi-formal construction lacks such continuity. The spatial identification of these building typologies enables governments and non-governmental organizations to focus technical assistance, material subsidies, and demonstration projects where they are likely to have the greatest protective effect.

The connection between buildings and foundations is another frequent point of failure during extreme wind events. Structures that are not securely anchored may slide, tilt, or collapse even when the walls and roofs remain intact. From a spatial perspective, foundation vulnerability often correlates with soil type, slope, floodplain location, and informal tenure status. Integrating geotechnical data with building inventories allows planners to anticipate compound failures in which wind, saturated soils, and poor anchoring interact.

The orientation and settlement pattern also shape wind exposure. Buildings aligned perpendicular to prevailing wind directions experience higher pressure loads on exposed façades. At the neighborhood scale, dense and irregular layouts may either amplify or dampen wind speeds depending on spacing and orientation. These effects are increasingly documented through post-event damage mapping and computational modeling, offering opportunities to incorporate wind-sensitive design principles into land-use planning and reconstruction.

8.4.3 Human Adaptation to Wind as a Cultural Practice

Long before formal hazard modeling, communities living in cyclone-prone regions developed adaptive strategies rooted in local knowledge, material availability, and collective memory. Vernacular architecture in the Caribbean, South Asia, and the western Pacific often reflects an understanding of wind behavior through low-profile structures, hipped roofs, flexible materials, and sacrificial elements designed to fail without triggering total collapse.

In many coastal and island societies, social organization itself functions as a wind adaptation mechanism. Extended family networks, communal shelters, shared labor for roof repair, and seasonal migration patterns reduce individual exposure to extreme events. These practices are not merely cultural artifacts. They are adaptive responses shaped by repeated encounters with cyclones. Geospatial documentation of such practices through participatory mapping, settlement history analysis, and post-event surveys can reveal forms of resilience that remain invisible in conventional hazard models.

Rapid urbanization, land scarcity, and standardized housing programs often disrupt these adaptive systems. Informal settlements may be pushed into exposed locations, while formal reconstruction efforts may overlook local wind dynamics. Recognizing and mapping the spatial footprint of traditional adaptation strategies helps bridge the gap between engineering-based resilience and lived experience.

8.4.4 Post-Disaster Assessment and Transparent Recovery

After a cyclone, rapid damage assessment is essential for directing aid, restoring services, and documenting losses. Satellite imagery, aerial photography, and crowd-sourced data enable near-real-time mapping of roof loss, vegetation damage, and infrastructure disruption. When combined with pre-storm exposure maps, these data support more equitable recovery by identifying where damage is most severe relative to population and need.

Transparency is a critical but often overlooked dimension of recovery. Maps that document damage, reconstruction progress, and resource allocation provide a public record that can support accountability and reduce corruption. In contexts where institutional trust is low, open geospatial platforms help bridge information gaps between governments, donors, and affected communities.

8.4.5 Looking Ahead: Wind Risk in a Warming World

Climate change is altering the wind regimes that cities have historically planned around. While the total number of tropical cyclones may not increase uniformly, there is strong evidence that the most intense storms are becoming stronger, wetter, and more destructive. Slower-moving storms extend the duration of damaging winds, increasing cumulative stress on buildings and infrastructure.

For cities already operating near the limits of housing quality, infrastructure capacity, and governance, these changes compress the margin for error. Geospatial tools cannot eliminate wind risk, but they can make it legible by revealing where adaptation investments are most likely to save lives, protect livelihoods, and reduce long-term recovery costs.

Emerging technologies, including machine-learning-based damage detection and real-time sensor networks, offer new possibilities for wind-risk analysis. Their effectiveness, however, depends on institutional coordination, data sharing, and sustained capacity building. Regional meteorological partnerships, shared storm databases, and interoperable spatial standards remain essential for translating local analyses into regional resilience.

Ultimately, getting to action on wind risk requires recognizing cyclones not as isolated disasters, but as recurring tests of urban systems. In this way, mapping wind risk transforms cyclones from episodic disasters into spatially legible urban threats that are shaped by climate, terrain, and governance, and therefore amenable to foresight, planning, and deliberate intervention.

References

Arsenault, K. R., Shukla, S., Hazra, A., Getirana, A., McNally, A., Kumar, S. V., Koster, R. D., Peters-Lidard, C. D., Zaitchik, B. F., Badr, H., Jung, H. C., Narapusetty, B., Navari, M., Wang, S., Mocko, D. M., Funk, C., Harrison, L., Husak, G. J., Adoum, A., Galu, G., Magadzire, T., Roningen, J., Shaw, M., Eylander, J., Bergaoui, K., McDonnell, R. A., & Verdin, J. P. (2020). The NASA Hydrological Forecast System for Food and Water Security Applications. Bulletin of the American Meteorological Society, 101(7), E1007-E1025. https://doi.org/10.1175/BAMS-D-18-0264.1

Australian Bureau of Meteorology. (2016). The 2015–16 El Niño. Bureau of Meteorology. https://www.bom.gov.au/climate/enso/2015-16/

Balk, D., Deichmann, U., Yetman, G., Pozzi, F., Hay, S. I., & Nelson, A. (2006). Determining global population distribution: Methods, applications and data. Advances in Parasitology, *62*, 119–156. https://doi.org/10.1016/S0065-308X(05)62004-0

Blocken, B. (2014). 50 years of computational wind engineering: Past, present and future. *Journal of Wind Engineering and Industrial Aerodynamics*, *129*, 69–102. https://doi.org/10.1016/j.jweia.2014.03.008

Cutter, S. L., Barnes, L., Berry, M., Burton, C., Evans, E., Tate, E., & Webb, J. (2008). A place-based model for understanding community resilience to natural disasters. *Global Environmental Change*, *18*(4), 598–606. https://doi.org/10.1016/j.gloenvcha.2008.07.013

Emanuel, K. (1986). An air–sea interaction theory for tropical cyclones. Part I: Steady-state maintenance. *Journal of the Atmospheric Sciences*, *43*(6), 585–604. https://doi.org/10.1175/1520-0469(1986)043%3C0585:AASITF%3E2.0.CO;2

Emanuel, K. (2005). Increasing destructiveness of tropical cyclones over the past 30 years. *Nature*, 436, 686–688. https://doi.org/10.1038/nature03906

Federal Emergency Management Agency (FEMA). (2012). *Engineering principles and practices for retrofitting flood-prone residential structures* (FEMA P-259). U.S. Department of Homeland Security.

Government of Fiji. (2016). Tropical Cyclone Winston: Situation report no. 28 (20 March 2016). National Disaster Management Office (NDMO).

Gray, W. M. (1979). Hurricanes: Their formation, structure, and likely role in the tropical circulation. In D. B. Shaw (Ed.), *Meteorology over the tropical oceans* (pp. 155–218). Royal Meteorological Society.

Grimmond, C. S. B., & Oke, T. R. (1999). Aerodynamic properties of urban areas derived from analysis of surface form. *Journal of Applied Meteorology*, *38*(9), 1262–1292.

Holland, G. J. (1980). An analytic model of the wind and pressure profiles in hurricanes. Monthly Weather Review, *108*(8), 1212–1218. https://doi.org/10.1175/1520-0493(1980)108<1212:AAMOTW>2.0.CO;2

Holton, J. R., & Hakim, G. J. (2013). *An introduction to dynamic meteorology* (5th ed.). Academic Press.

IPCC. (2022). *Sixth assessment report: Working group II – Impacts, adaptation, and vulnerability*. Cambridge University Press.

Knutson, T. R., Camargo, S. J., Chan, J. C. L., Emanuel, K., Ho, C.-H., Kossin, J., Mohapatra, M., Satoh, M., Sugi, M., Walsh, K., & Wu, L. (2020). Tropical cyclones and climate change assessment: Part II. *Bulletin of the American Meteorological Society*, *101*(3), E303–E322. https://doi.org/10.1175/BAMS-D-18-0194.1

Kossin, J. P. (2018). A global slowdown of tropical-cyclone translation speed. *Nature*, 558, 104–107. https://doi.org/10.1038/s41586-018-0158-3

Landsea, C. W., & Franklin, J. L. (2013). Atlantic hurricane database uncertainty and presentation of a new database format (HURDAT2). Monthly Weather Review, *141*(10), 3576–3592. https://doi.org/10.1175/MWR-D-12-00254.1

Miller, C., Barlage, M., Zeng, X., & Wei, H. (2013). Sensitivity of land–atmosphere interactions to terrain representation. *Journal of Geophysical Research: Atmospheres*, 118(10), 1–15.

NOAA National Hurricane Center. (2024). *Atlantic hurricane database (HURDAT2)*. https://www.nhc.noaa.gov/data/

Oke, T. R. (1987). *Boundary layer climates* (2nd ed.). Routledge.

Paulay, T., & Priestley, M. J. N. (1992). Seismic design of reinforced concrete and masonry buildings. John Wiley & Sons.

Pelling, M. (2011). *Adaptation to climate change: From resilience to transformation*. Routledge.

Smith, R. B. (1979). The influence of mountains on the atmosphere. *Advances in Geophysics*, *21*, 87–230.

Tamura, Y., Kareem, A., Simiu, E., & Solari, G. (2015). *Wind effects on buildings and structures*. Springer.

Taubenböck, H., Wegmann, M., Roth, A., Mehl, H., & Dech, S. (2013). Urbanization in India – Spatiotemporal analysis using remote sensing data. Computers, Environment and Urban Systems, *38*, 39–52. https://doi.org/10.1016/j.compenvurbsys.2012.06.003

Trapp, R. J., Diffenbaugh, N. S., Brooks, H. E., Baldwin, M. E., Robinson, E. D., & Pal, J. S. (2016). Changes in severe thunderstorm environment frequency during the 21st century caused by anthropogenically enhanced global radiative forcing. Proceedings of the National Academy of Sciences, *104*(50), 19719–19723. https://doi.org/10.1073/pnas.0705494104

UNDRR. (2015). *Sendai framework for disaster risk reduction 2015–2030*. United Nations Office for Disaster Risk Reduction.

Vickery, P. J., Skerlj, P. F., Lin, J., Twisdale, L. A., Young, M. A., & Lavelle, F. M. (2006). HAZUS-MH hurricane model methodology. II: Damage and loss estimation. Natural Hazards Review, *7*(2), 94–103. https://doi.org/10.1061/(ASCE)1527-6988(2006)7:2(94)

World Meteorological Organization (WMO). (2017). *Tropical cyclone programme: Annual report*. WMO-No. 1212.

9

Mapping Landslides

Chapter Introduction

Landslides are often described as sudden natural disasters, but in urban settings, they are better understood as the outcome of long-term processes unfolding across terrain, infrastructure, and governance systems. Slope instability emerges from the interaction of geology, hydrology, vegetation, and land use, yet whether a landslide becomes a disaster depends on where people live, how land is managed, and which risks are tolerated or ignored over time.

In many cities, especially those expanding rapidly into steep or marginal terrain, landslide risk accumulates quietly. Informal housing, road construction, deforestation, drainage modification, and inadequate enforcement of land-use regulations all alter slope stability incrementally. Heavy rainfall or seismic activity may trigger failure, but these events act on landscapes that have already been made vulnerable through human decisions.

Understanding landslides as spatially produced risks rather than isolated events is essential for effective mapping and intervention. Geospatial analysis allows these interacting factors to be examined together, revealing patterns of susceptibility and exposure that are not visible through hazard inventories alone. This framing shifts attention from post-disaster response toward prevention, regulation, and risk-informed urban development.

9.1 Setting the Scene: Regent, Freetown

In the early hours of August 14, 2017, a catastrophic landslide swept through the Regent neighborhood on the northern slope of Sugar Loaf Mountain above Freetown, Sierra Leone, killing more than a thousand people and displacing many thousands more. Triggered after days of intense monsoonal rainfall, the failure began high on the massif and rapidly transformed into a fast-moving debris flow that overwhelmed houses, roads, and drainage channels downslope. Entire blocks were buried under meters of mud, boulders, and uprooted vegetation. The scale of the destruction prompted international headlines and humanitarian response, but the event itself was not

DOI: 10.1201/9781032636948-9

an anomaly. Rather, it was the tragic culmination of rapid urban growth, changes in land cover, and poor planning.

Freetown's physical setting has always posed challenges for urban settlement. The city is wedged between the Atlantic Ocean and a steep interior escarpment formed by weathered volcanic and granitic bedrock. Soils in this region are deeply leached by a humid tropical climate characterized by intense seasonal rainfall. These lateritic and saprolitic soils are typically nutrient-poor, highly porous, and structurally weak when saturated. Under forest cover, deep root systems and organic matter can impart cohesion and regulate infiltration. Once vegetation is removed, however, these soils lose much of their mechanical strength and become highly susceptible to mass wasting under prolonged rainfall.

The Regent neighborhood exemplifies this vulnerability. Originally covered in tropical woodland, the hillsides above Freetown experienced accelerating deforestation over several decades as the city expanded upslope. Formal planning mechanisms struggled to keep pace with population growth, and informal housing proliferated along steep gradients ill-suited for construction. Houses were often built incrementally, without engineered foundations or adequate drainage. As tree cover declined, rainfall infiltration increased, groundwater tables rose, and slope stability margins narrowed. These changes were gradual and largely invisible, accumulating silently until a threshold was crossed. In Figure 9.1, a girl looks at the mountainside

FIGURE 9.1
Regent landslide aftermath. (Photo taken after the devastating landslide hit on August 14, 2017. Photo: Mark Stedman. This file is licensed under the Creative Commons Attribution 2.0 Generic license.)

that gave way in Freetown, Sierra Leone. Her family home and many others were destroyed when a devastating landslide hit on August 14, 2017.

Climate change intensified these underlying risks. While landslides have long occurred in Sierra Leone's highlands, the frequency and intensity of extreme rainfall events have increased across West Africa. In August 2017, rainfall totals exceeded seasonal norms, saturating soils over multiple days. Without roots to bind the soil matrix or drainage infrastructure to divert runoff, gravitational forces overwhelmed resisting forces. The landslide was not sudden in a scientific sense; it was the predictable outcome of hydrological loading acting on a destabilized slope.

Political and institutional factors further shaped the disaster's impact. At the time of the landslide, Freetown was undergoing a period of governance transition, with limited enforcement of land-use controls in peri-urban areas. Hazard mapping existed in fragmented forms but was not systematically integrated into development decisions or community awareness. Residents often understood the risks abstractly yet lacked viable alternatives for housing in a city where low-lying areas were already densely occupied and flood prone.

The Regent landslide illustrates a central argument of this chapter: landslides are rarely "acts of God." They are the outcome of identifiable interactions among geology, climate, land cover, and urbanization. GIS provides the tools to make these interactions visible before disaster strikes—by mapping slope, soil properties, vegetation loss, and settlement patterns together. In cities like Freetown, where steep terrain intersects with rapid urban growth, the challenge is not a lack of scientific understanding, but the failure to translate spatial knowledge into preventive action.

Landslides are often seen as sudden and unpredictable, but most result from predictable combinations of geology, slope, rainfall, and human activity. In steep, densely populated cities, deforestation and informal construction on unstable hillsides magnify the danger. Using short case studies from Puerto Rico and Manizales, Colombia, this chapter explores how GIS can reveal slope instability and guide safer land-use planning.

9.2 Unpacking the Threat: How and Why Landslides Occur

Landslides belong to a broader class of geomorphic processes known as mass wasting: the downslope movement of soil, rock, or debris under the direct influence of gravity. Although often described as sudden or catastrophic, landslides are more accurately understood as the endpoint of cumulative processes that progressively erode slope stability over time (Cruden & Varnes, 1996). From a geospatial perspective, landslides occur where gravitational driving forces exceed the resisting forces within a slope system.

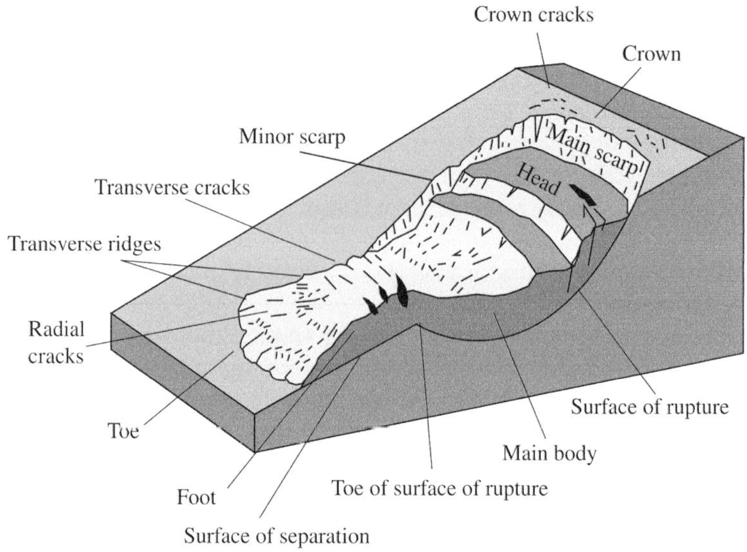

FIGURE 9.2
Slope failure illustrated. (Source: U.S. Geological Survey landslide process models.)

This imbalance emerges from spatially patterned changes in slope geometry, material properties, hydrological conditions, and external disturbances, all of which can be examined through mapping and analysis (Guzzetti et al., 1999; Van Westen et al., 2008).

Figure 9.2 illustrates the functional elements of a landslide. As rainfall infiltrates hillslope materials, soil moisture increases and raises pore-water pressure within a developing saturated zone. As effective stress and shear strength decline, failure occurs along a slip surface, resulting in downslope movement of the failed mass.

Slope angle exerts a primary control on landslide susceptibility. Steeper slopes experience higher shear stress due to gravity, bringing them closer to failure even under undisturbed conditions (Montgomery & Dietrich, 1994). Yet slope alone does not explain landslide occurrence. Many steep slopes remain stable for long periods, while others fail repeatedly. The difference lies in the interaction between topography and material properties. Soil depth, grain size distribution, mineral composition, and degree of weathering all influence shear strength (Sidle & Ochiai, 2006). In tropical environments, prolonged chemical weathering produces thick regolith layers that may appear stable when dry but lose cohesion rapidly once saturated (Thomas, 1994; Glade, 2003).

Hydrology is often the decisive triggering mechanism. Intense rainfall or prolonged wet periods increase pore-water pressure within soils, reducing frictional resistance along potential failure planes (Iverson, 2000).

Antecedent moisture conditions are as important as rainfall intensity itself; a moderate storm following weeks of saturation can be more destabilizing than a single extreme downpour (Crozier, 1999; Bogaard & Greco, 2018). In monsoonal climates, this cumulative effect is especially pronounced. GIS-based rainfall surfaces, soil-moisture proxies, and terrain indices such as the Topographic Wetness Index provide spatial approximations of these processes, even where direct measurements are sparse (Beven & Kirkby, 1979; Moore et al., 1991).

Vegetation plays a dual role in slope stability. Root systems mechanically reinforce soils by increasing cohesion, while canopy cover reduces surface erosion and moderates infiltration (Wu et al., 1979; Greenway, 1987). Vegetation removal—through logging, agriculture, road construction, or informal settlement—weakens both stabilizing functions (Sidle & Ochiai, 2006). In urban contexts, this loss is rarely uniform. It occurs in patches following development fronts, access routes, and property boundaries, producing localized zones of heightened susceptibility. These zones are often invisible in coarse regional hazard maps but become apparent when land-cover change is analyzed at finer spatial and temporal scales (Turner et al., 2003; Van Westen et al., 2008).

Human modification of slopes frequently accelerates destabilization. Excavation for building pads, road cuts, and utilities alters slope geometry and removes lateral support, increasing shear stress on already marginal terrain (Crozier, 2010). Poorly designed drainage concentrates runoff rather than dispersing it, increasing erosion and saturation downslope (Montgomery et al., 2000). In informal settlements, such interventions are rarely coordinated or engineered, producing cumulative effects that far exceed the impact of any single structure. GIS allows these incremental changes to be reconstructed over time through comparisons of historical imagery, elevation models, and land-use maps, revealing patterns of risk accumulation rather than isolated failures (Pelletier et al., 2015).

Seismic activity adds another layer of complexity. Earthquakes can directly trigger landslides by rapidly increasing shear stress or liquefying saturated soils, but their influence extends well beyond the moment of shaking (Keefer, 1984; Keefer, 2002). Seismic regions often exhibit fractured bedrock and weakened soil profiles that remain susceptible to rainfall-induced landslides long after an earthquake has passed (Fan et al., 2019). Importantly, seismic risk is not confined to major plate boundaries. Intraplate earthquakes occur in many regions assumed to be stable, reflecting broader patterns in which societies accept geophysical risk where economic opportunity or political inertia outweigh precautionary planning (Vale & Campanella, 2005).

The scientific concept integrating these variables is the factor of safety, defined as the ratio of resisting forces to driving forces acting on a slope (Duncan & Wright, 2005). When this ratio falls below one, failure becomes inevitable. While originating in geotechnical engineering, the logic of the factor of safety underpins many GIS-based landslide susceptibility models. Spatial layers representing slope, lithology, land cover, rainfall, and seismicity function as proxies for its components (Guzzetti et al., 1999; Reichenbach et al., 2018). The objective is not to calculate precise thresholds everywhere, but to identify areas where safety margins are systematically eroded by environmental and human pressures.

Urbanization introduces a critical social dimension to landslide risk. In many cities in the Global South, residential sorting pushes lower-income populations onto steep hillsides, unstable embankments, and poorly serviced peri-urban slopes (Davis, 2006; UN-Habitat, 2018). This pattern is not accidental. It reflects land markets, governance failures, and historical inequalities that concentrate vulnerability in ecologically marginal spaces (Anguelovski et al., 2016). Deforestation on slopes is therefore not merely an environmental problem, but a spatial expression of social inequality.

Climate change amplifies these vulnerabilities rather than creating entirely new ones. The intensification of the hydrological cycle increases the frequency and severity of extreme precipitation events in many regions (IPCC, 2022). In coastal and estuarine cities, sea-level rise can further slow drainage, increasing upstream saturation during storms and indirectly heightening landslide risk (Nicholls et al., 2007; Wong et al., 2014). In mountainous coastal cities, landslides, floods, and debris flows often occur together, challenging planning frameworks that treat hazards as separate phenomena (Gill & Malamud, 2014).

The impacts of landslides extend well beyond immediate mortality and physical destruction. Infrastructure damage can sever transport corridors, disrupt water supply, and isolate neighborhoods for prolonged periods (Petley, 2012). Economic losses fall disproportionately on households with limited insurance or access to formal credit, particularly in informal settlements (World Bank, 2017). Environmental effects include vegetation loss, soil degradation, and altered drainage networks that increase susceptibility to future events (Restrepo et al., 2009). These cascading consequences underscore why landslides must be understood as chronic urban risks rather than episodic disasters.

Taken together, these dynamics challenge the framing of landslides as unpredictable natural events. While the precise timing of failure may be uncertain, the conditions that produce landslides are neither random nor unknowable. They are spatially structured, historically contingent, and deeply intertwined with urban development choices (Alexander, 2005; UNDRR, 2022). GIS provides a means of synthesizing these dimensions, translating complex earth-system processes into actionable spatial knowledge. The next section turns from understanding landslide mechanisms to modeling risk, showing how geospatial tools can expose hidden vulnerabilities and support preventive intervention rather than post-disaster explanation.

9.3 Modeling Landslide Risk in Geospatial Terms

Modeling landslide risk in geospatial terms requires translating geomorphic processes, hydrological conditions, and human interventions into spatial representations that can be evaluated systematically across landscapes. Unlike hazards with clearly bounded footprints, landslide susceptibility varies continuously across space and time, shaped by slope geometry, material properties, land cover, and triggering conditions. As a result, landslide modeling focuses less on predicting individual failures and more on identifying zones where stability margins are persistently low and where small disturbances are likely to produce disproportionate impacts.

Most geospatial landslide models combine terrain-derived variables with environmental and anthropogenic indicators to approximate slope stability. Digital Elevation Models (DEMs) provide the basis for calculating slope, curvature, aspect, and contributing area, while additional layers such as lithology, land cover, rainfall intensity, and proximity to roads or development capture key drivers of destabilization (Van Westen et al. 2008; Guzzetti et al. 1999). These inputs are integrated using heuristic, statistical, or physically based approaches to generate susceptibility surfaces that rank relative risk across space rather than deterministic predictions of failure.

DEMs form the backbone of most landslide analyses. From DEMs, analysts derive slope angle, aspect, curvature, flow accumulation, and indices of potential soil saturation. Slope angle captures the gravitational component of instability, while curvature helps distinguish convex ridges from concave hollows where water and debris tend to accumulate (Montgomery & Dietrich, 1994). Aspect influences microclimate and vegetation patterns, affecting soil moisture and root strength. Hydrologically derived metrics such as the Topographic Wetness Index and stream power index serve as proxies for subsurface water concentration and erosive potential, both of which are critical to understanding failure mechanisms (Beven & Kirkby, 1979; Moore et al., 1991).

Terrain alone, however, cannot explain landslide patterns. Geological and soil layers provide essential context by representing the material properties of slopes. Lithology influences weathering rates, fracture density, and shear strength, while soil texture and depth affect permeability and cohesion (Sidle & Ochiai, 2006). In many regions of the Global South, detailed geotechnical data are sparse or unevenly distributed. GIS analysts therefore rely on generalized geological maps, global soil datasets, or terrain-based proxies. While these substitutions introduce uncertainty, they nonetheless improve upon models that consider topography in isolation (Van Westen et al., 2008).

Land cover and land-use change introduce a critical human dimension to landslide modeling. Vegetation loss, impervious surface expansion, and slope cutting alter hydrological pathways and mechanical stability in ways that are spatially heterogeneous. Time-series satellite imagery allows analysts to reconstruct patterns of deforestation and urban expansion, linking recent land-cover change to emerging instability (Turner et al., 2003; Hansen et al., 2013). In rapidly urbanizing cities, these changes often occur faster than official maps are updated, making remote sensing an indispensable complement to administrative data.

Rainfall acts as the most common trigger for landslides worldwide, particularly in humid tropical and monsoonal climates. GIS-based models incorporate rainfall either indirectly, through long-term climatic averages, or directly, through event-based precipitation surfaces derived from gauges or satellites (Hong et al., 2006). Increasingly, models distinguish between short-duration high-intensity rainfall and prolonged antecedent wetness, reflecting advances in hydrological understanding of slope failure thresholds (Crozier, 1999; Bogaard & Greco, 2018). While precise triggering thresholds are difficult to generalize across regions, spatial rainfall data allow analysts to identify areas where slopes are repeatedly exposed to destabilizing moisture regimes.

Seismicity adds another layer of complexity. Earthquake-induced landslides can be modeled by integrating fault maps, peak ground acceleration surfaces, and historical shaking intensity data (Keefer, 2002; Tanyaş et al., 2017). Even in regions where earthquakes are infrequent, the inclusion of seismic layers helps identify slopes that may have been structurally weakened by past events. Importantly, seismic risk layers often overlap with dense urban development, underscoring how economic and political decisions concentrate populations in hazardous zones rather than dispersing them.

The integration of these variables reflects a broader conceptual shift from hazard mapping to risk modeling. Hazard layers describe where landslides are likely to occur; risk emerges only when these hazards intersect with exposed populations, infrastructure, and social vulnerability (UNDRR, 2022). GIS enables this integration by allowing physical susceptibility surfaces to be combined with demographic data, building footprints, transportation networks, and critical facilities. In doing so, landslide modeling moves beyond geomorphology into the realm of urban governance, revealing how risk is produced through land-use decisions and uneven access to safety.

Different modeling approaches embody different assumptions about causality and uncertainty. Heuristic models weight input layers based on expert judgment, while statistical approaches use historical landslide inventories to infer relationships between conditioning factors and observed failures (Guzzetti et al., 1999; Chung & Fabbri, 2003). Physically based models attempt to approximate slope stability processes more directly, often through simplified representations of the factor of safety across space (Montgomery & Dietrich, 1994; Iverson, 2000). Each approach involves trade-offs between data requirements, interpretability, and transferability, particularly when applied in data-poor urban contexts.

What unites these approaches is their capacity to make latent risk visible. Landslide models do not predict the exact moment of failure, nor do they eliminate uncertainty. Instead, they identify spatial patterns of heightened susceptibility that would otherwise remain obscured until disaster strikes. In cities like Freetown or Manizales, this visibility is itself a form of intervention. By showing where development has encroached onto unstable slopes, GIS challenges narratives of inevitability and reframes landslides as outcomes of knowable—and therefore governable—processes.

In the sections that follow, this chapter turns from conceptual foundations to applied practice. It examines how open-source data and widely available tools can be used to construct landslide susceptibility maps, how more advanced models incorporate social vulnerability and institutional factors, and how these analyses can inform land-use regulation, early warning systems, and community-based risk reduction. Modeling landslides in GIS is not simply an academic exercise; it is a means of confronting the political and spatial choices that place cities in harm's way.

9.3.1 Project Design and Applied Modeling

Translating conceptual models into applied analysis requires careful attention to project design. Landslide modeling begins not with software, but with a set of explicit analytical questions: What type of mass wasting is of concern? Is the objective to understand chronic susceptibility, to assess exposure under current land-use patterns, or to evaluate potential change under alternative development scenarios? These questions shape decisions about spatial scale, variable selection, and acceptable uncertainty. In urban settings, especially in the Global South, analysts must often work with imperfect data, making transparency about assumptions as important as technical sophistication (Van Westen et al., 2008; Alexander, 2008).

A typical GIS workflow for landslide susceptibility mapping integrates several core datasets. Elevation data provide the structural framework through DEM-derived slope, aspect, and curvature surfaces. Global DEMs at 30-meter resolution are widely available, while higher-resolution products—often derived from LiDAR—allow much finer representation of micro-topography where data exist (Gesch et al., 2002). Geological and soil datasets introduce material properties, even when generalized, and land-cover layers capture vegetation condition and urban encroachment. These layers are standardized to a common spatial resolution and extent, then combined using either weighted overlays or statistical relationships calibrated against known landslide locations (Guzzetti et al., 2005; Reichenbach et al., 2018).

Rainfall data are often incorporated as dynamic inputs rather than static layers. Satellite-based precipitation products allow analysts to examine spatial variability in storm intensity and duration, particularly in regions with sparse gauge networks (Hong et al., 2006). When historical landslide inventories are available, rainfall thresholds can be empirically linked to observed failures, improving confidence in susceptibility estimates (Crozier, 1999). In many cities, however, such inventories are incomplete or inconsistent.

In these cases, modeling emphasizes relative risk rather than absolute prediction, highlighting zones of concern rather than attempting deterministic forecasts.

More advanced modeling approaches seek to approximate the physical processes underlying slope failure more directly. Physically based models estimate spatial variations in the factor of safety by combining slope geometry, soil strength parameters, and hydrological loading (Montgomery & Dietrich, 1994; Iverson, 2000). While these models offer conceptual clarity, especially in the Global South, they require detailed input data that are rarely available at city scale. As a result, they are often applied selectively, such as along critical infrastructure corridors or in high-risk neighborhoods identified through simpler screening analyses. GIS enables this tiered approach, allowing coarse models to guide where more intensive investigation is warranted.

Seismic risk modeling introduces additional challenges. Earthquake-induced landslide susceptibility depends on ground shaking intensity, slope conditions, and material properties, all of which vary spatially and temporally. Peak ground acceleration maps and fault datasets provide a starting point, but translating seismic hazard into landslide risk requires assumptions about how shaking interacts with saturated soils and fractured bedrock (Keefer, 2002; Tanyaş et al., 2017). In many cities, seismic layers are included not to predict imminent failure, but to identify compound hazards—areas where earthquakes, rainfall, and unstable slopes coincide, magnifying potential impacts.

A growing body of work integrates social vulnerability into landslide risk modeling. Physical susceptibility alone cannot explain why landslides become disasters in some places and not others. GIS allows demographic data—income, housing quality, age structure, tenure status—to be overlaid with hazard surfaces, revealing uneven exposure and capacity to cope (Cutter et al., 2003; Birkmann et al., 2013). Composite indices such as the Social Vulnerability Index (SoVI) or Global Vulnerability Index (GVI) operationalize these dimensions, though their construction involves normative choices about which variables matter most (Cutter & Finch, 2008). When applied carefully, these indices shift attention from slopes alone to the populations living upon them.

Remote sensing has expanded the temporal dimension of landslide analysis. Pre- and post-event DEM differencing can quantify the volume of material moved during a landslide, providing insight into process magnitude and downstream impacts (Coe et al., 1997). Time-series imagery reveals slow-moving slope deformation that may precede catastrophic failure, blurring the line between acute disaster and chronic instability (Handwerger et al., 2019). These techniques are particularly valuable for monitoring urban hillsides where gradual movement may go unnoticed until thresholds are exceeded.

Economic and institutional factors shape how landslide risk is modeled and acted upon. Governments may possess hazard maps but lack the legal authority or political will to restrict development in high-risk zones, though this may depend on the level of governance. In other cases, zoning regulations exist on paper but are routinely circumvented through informal construction or selective enforcement (Hardoy & Satterthwaite, 2014). Insurance markets, where they exist at all, often exclude landslide risk or impose conditions that effectively bar low-income households from coverage (World Bank, 2017). These institutional realities influence not only exposure but also the types of data that are collected and the models that are prioritized.

The absence of early warning systems in many landslide-prone cities reflects these same constraints. Technically, low-cost monitoring systems—combining rainfall sensors, ground movement detectors, and community reporting—are increasingly feasible (Intrieri et al., 2013). The challenge lies less in technology than in governance: sustaining maintenance, ensuring trust in warnings, and integrating alerts into decision-making processes. GIS plays a role here by linking sensor data to spatial thresholds, translating abstract measurements into place-based warnings that residents and officials can act upon.

Ultimately, landslide modeling in GIS is as much a political exercise as a technical one. Choices about scale, variables, and thresholds encode assumptions about what risks are tolerable and whose lives are valued. By making these choices explicit, geospatial analysis opens space for debate and accountability. In cities confronting rapid growth on unstable terrain, the value of modeling lies not in prediction alone, but in its capacity to reframe landslides as foreseeable consequences of development pathways—and thus as hazards that can be mitigated through informed planning rather than endured as inevitable losses.

Case Study 9.1: A Bivariate Approach to Physical and Social Landslide Risk: Puerto Rico

To illustrate how physical susceptibility and social vulnerability can be integrated in a single analytical framework, this chapter draws on a landslide risk assessment conducted in Puerto Rico following Hurricane Maria in 2017. The analysis was refocused on Puerto Rico not for conceptual reasons, but because of unusually strong data availability relative to many other landslide-prone regions in the tropics. High-resolution terrain data, standardized demographic datasets, and pre-existing landslide susceptibility products make it possible to demonstrate a transparent, replicable methodology linking geophysical conditions to population-level vulnerability.

The analysis concentrated on the municipality of Utuado, one of the areas most severely affected by mass wasting events triggered by extreme rainfall during and after Maria. Utuado's steep terrain, deeply weathered soils, and dispersed settlement pattern made it an appropriate case for examining how landslide risk emerges from the interaction of slope processes and social conditions rather than from either dimension alone (Larsen & Torres-Sánchez, 1998; Restrepo et al., 2009). Rather than attempting to predict individual landslide events, the goal was to map relative risk across the municipality in a way that could inform planning and post-disaster recovery.

The geospatial workflow combined multiple datasets commonly used in landslide and vulnerability studies. Census TIGER boundaries provided a consistent spatial framework for demographic analysis, while the SVI was joined to these units to represent differential capacity to prepare for, respond to, and recover from hazard impacts (Cutter et al., 2003; Cutter & Finch, 2008). Physical susceptibility was represented using soil data and an existing landslide susceptibility surface derived from terrain and hydrological variables. These layers captured the likelihood of slope failure based on environmental conditions, independent of who occupied the land.

Following the approach developed by Azar and Rain (2009) in their analysis of landslide risk in Puerto Rico, physical and social components were first evaluated separately, then combined into a composite risk surface. Rather than collapsing variables into a single index prematurely, the method retained the distinction between biophysical and social vulnerability, allowing each to be mapped and interpreted on its own terms.

As shown in Figure 9.3, using open data, it is possible to construct a bivariate map in which one axis represents physical landslide susceptibility and the other represents social vulnerability. This approach can easily make visible areas where high geophysical risk coincides with populations such as elderly housing or hospitals that are least able to cope with disruption.

The value of this bivariate framework lies less in statistical sophistication than in interpretability. By preserving the dual structure of risk, the

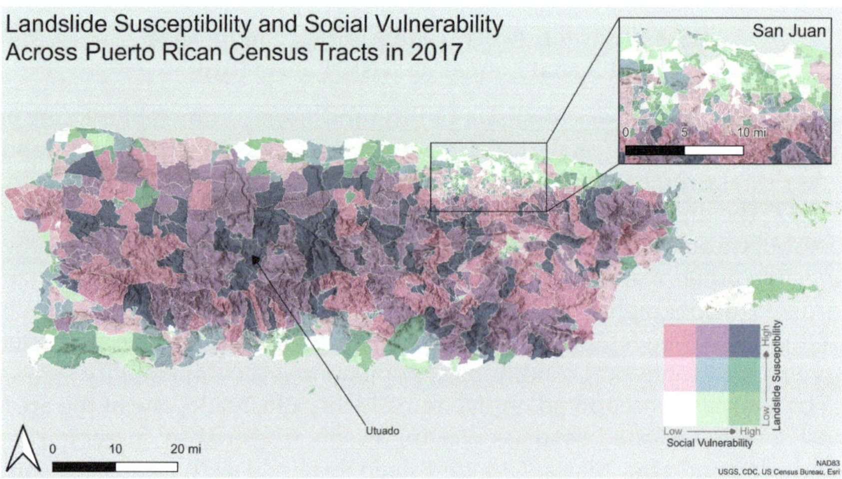

FIGURE 9.3
Bivariate landslide risk and SoVI (Social Vulnerability Index) Map. (Cartography by John Lesko.)

map avoids the false precision of composite scores while still supporting decision-making. Areas of high susceptibility but low vulnerability suggest priorities for engineering intervention or land-use regulation, while areas where both dimensions are high point toward urgent needs for relocation assistance, infrastructure investment, or community-based mitigation. In this sense, the Puerto Rico example functions as a methodological template rather than a site-specific solution, demonstrating how GIS can be used to integrate physical and social dimensions of landslide risk in data-rich contexts—and, by extension, how similar logic might be adapted in cities where data are less complete, but the stakes are equally high.

9.4 Getting to Action on Landslide Risk

Reducing landslide risk requires moving from descriptive hazard mapping to interventions that reshape how cities interact with unstable terrain. Because landslides are closely tied to land use, drainage, and settlement patterns, effective action depends on anticipatory planning rather than post-event response. Geospatial analysis supports this shift by revealing where slope instability intersects with population exposure, infrastructure networks, and institutional capacity, allowing cities to intervene before failure occurs.

9.4.1 From Susceptibility Mapping to Land-Use Control

One of the most direct applications of landslide modeling is land-use regulation. Susceptibility maps identify slopes where development substantially increases the likelihood of failure, providing an empirical basis for zoning restrictions, setback requirements, and development moratoria. When incorporated into permitting systems, these maps help prevent incremental encroachment onto unstable hillsides that gradually erodes safety margins.

In many Global South cities, however, land-use controls are weak or unevenly enforced. Informal settlement often expands onto steep slopes precisely because safer land is inaccessible or unaffordable. In these contexts, landslide mapping can still guide action by identifying priority areas for upgrading, relocation assistance, or infrastructure reinforcement rather than serving solely as exclusionary tools.

9.4.2 Drainage, Infrastructure, and Slope Stabilization

Hydrological management is among the most effective interventions for reducing landslide risk. Because saturation and pore-water pressure play central roles in triggering failure, improvements to surface drainage, subsurface conveyance, and erosion control can significantly enhance slope stability. GIS-based analyses help identify where runoff is concentrated, where drainage networks are interrupted, and where informal infrastructure channels water onto vulnerable slopes.

Infrastructure design decisions also shape landslide outcomes. Roads, pipelines, and retaining structures frequently cut across slopes, altering load distributions and concentrating stress. Mapping the spatial relationship between infrastructure corridors and historical landslides helps planners identify high-risk alignments and prioritize retrofits. In some cases, relatively modest interventions—such as improved culverts, slope regrading, or vegetation restoration—can substantially reduce failure probability.

9.4.3 Community-Based Monitoring and Early Warning

In many landslide-prone cities, formal monitoring networks are sparse or nonexistent. Community-based observation therefore plays a critical role in detecting early signs of instability, including ground cracks, leaning trees, blocked drains, or changes in spring flow. When combined with geospatial platforms, these observations can be geolocated, aggregated, and linked to rainfall thresholds to support early warning systems.

Participatory mapping initiatives allow residents to document local knowledge of slope behavior and past failures that may not appear in official records. These efforts not only improve data quality but also strengthen risk awareness and trust between communities and authorities. Importantly, early warning systems are most effective when paired with clear evacuation protocols and credible response capacity.

9.4.4 Recovery, Relocation, and Long-Term Risk Reduction

Post-landslide recovery presents both challenges and opportunities. Emergency reconstruction often reproduces pre-existing risk by rebuilding in place without addressing underlying instability. Geospatial damage assessments can support more strategic recovery by identifying areas where reconstruction is unsafe and where relocation may offer a more durable solution.

Relocation is socially and politically sensitive, particularly where livelihoods and social networks are tied to specific locations. Spatial analysis can help identify alternative sites that balance safety, accessibility, and economic opportunity, reducing the likelihood that displaced populations simply move from one hazard zone to another. In this way, mapping landslide risk transforms slope failure from an unpredictable act of nature into a spatially legible urban threat—shaped by terrain, hydrology, and development choices, and therefore open to prevention through deliberate, place-based planning.

COMMUNITY-LED LANDSLIDE RISK REDUCTION: GUARDIANS OF THE SLOPE, MANIZALES, COLOMBIA

The "*Guardianas de la Ladera*" (Guardians of the Slope) program in Manizales, Colombia, provides a widely cited example of how community-based initiatives can be embedded within formal urban risk management systems. Launched in 2003, the program employs women—primarily heads of household from vulnerable communities—to maintain slope stabilization infrastructure in landslide-prone neighborhoods (Mejía, 2006; Wesely, 2021). Manizales' steep topography, high annual rainfall, volcanic soils, and seismic exposure make landslides a persistent urban threat, particularly in informal settlements that expanded rapidly during the late twentieth century (Velásquez Barrero, 1998). This section draws on background research compiled by Anke Marais as part of a supervised independent study conducted under my direction in the Master's program in International Development Studies in GW's Elliott School.

The Guardians are responsible for monitoring and maintaining more than a thousand stabilization works, including drainage channels, retaining walls, and vegetated slopes. Their daily activities include clearing debris, identifying early signs of erosion or structural failure, and reporting emerging risks to municipal authorities. By performing routine maintenance that would otherwise overwhelm city agencies, the program reduces the likelihood that minor problems escalate into catastrophic slope failures.

Beyond its technical function, the program has significant social impacts. By providing paid employment and technical training to women from marginalized communities, the Guardians initiative strengthens household economic security while elevating women's roles in neighborhood governance (Wesely, 2021). As visible stewards of risk reduction, the Guardians also act as trusted intermediaries between residents and municipal institutions, reinforcing a culture of prevention and shared responsibility.

The Guardians of the Slope have been widely recognized as a "good practice" in urban disaster risk reduction by international agencies, including the United Nations Development Programme (Mejía Prieto et al., 2006). Their success illustrates how landslide mitigation can generate co-benefits: improved physical safety, strengthened social cohesion, and incremental formalization of informal settlements. At the same time, the program's limits are instructive. Community-based maintenance cannot substitute for major engineering works or resolve structural inequalities on its own. Its effectiveness depends on sustained political support, technical oversight, and integration into broader land-use and environmental policies. Recent reporting from the Colombian state notes that the program is going strong and was expanded to other areas in 2024. The Guardians of the Slope model demonstrates that meaningful action on landslide risk is possible even in resource-constrained settings when communities are treated as partners rather than passive beneficiaries (Marais, 2025).

A contrasting example from the European Alps underscores that landslides and avalanches are not inherently catastrophic when embedded within robust monitoring, governance, and preparedness systems. In May 2025, the Swiss village of Blatten was almost entirely destroyed by an avalanche triggered by the collapse of the Birch Glacier, which released ice, rock, and debris into the valley below. Despite the near-total physical devastation, loss of life was avoided because residents had been evacuated in advance following warnings based on continuous slope and glacier monitoring (Swiss Federal Office for the Environment, 2025; Guardian, 2025). The Blatten case illustrates that even in environments where geomorphic hazards cannot be prevented, outcomes are shaped decisively by early warning, institutional trust, and the political legitimacy of restrictive land-use decisions. The comparison is instructive: While the physical processes driving avalanches in Switzerland differ from rainfall-induced landslides in tropical cities, the underlying lesson is the same. Disaster severity reflects not only environmental conditions but also the extent to which societies invest in anticipating failure, enforcing precaution, and valuing collective safety over short-term development gains.

References

Alexander, D. (2005). Towards the development of standards in emergency management training and education. *Disaster Prevention and Management*, *14*(2), 158–175.

Alexander, D. (2008). Scenario methodology for teaching principles of emergency management. *Disaster Prevention and Management*, *17*(2), 171–180.

Anguelovski, I., Connolly, J. J. T., Masip, L., & Pearsall, H. (2016). Assessing green gentrification in historically disenfranchised neighborhoods. *Urban Geography*, *39*(3), 458–491.

Azar, D., & Rain, D. (2009). Populations at risk: Social vulnerability and landslide hazards in Puerto Rico. *Geographical Review*, *99*(4), 451–472.

Beven, K. J., & Kirkby, M. J. (1979). A physically based, variable contributing area model of basin hydrology. *Hydrological Sciences Bulletin*, *24*(1), 43–69.

Birkmann, J., Cardona, O. D., Carreno, M. L., et al. (2013). Framing vulnerability, risk and societal responses. *Natural Hazards*, *67*(2), 193–211.

Bogaard, T. A., & Greco, R. (2018). Landslide hydrology: From hydrology to pore pressure. *Wiley Interdisciplinary Reviews: Water*, *5*(4), e1291.
Chung, C. F., & Fabbri, A. G. (2003). Validation of spatial prediction models for landslide hazard mapping. *Natural Hazards*, *30*(3), 451–472.
Coe, J. A., Glancy, P. A., & Whitney, J. W. (1997). Volume estimation of a landslide from repeated surveys. *Engineering Geology*, *46*(1), 151–162.
Crozier, M. J. (1999). Prediction of rainfall-triggered landslides. *Earth Surface Processes and Landforms*, *24*(9), 825–834.
Crozier, M. J. (2010). Deciphering the effect of climate change on landslide activity. *Geomorphology*, *124*(3–4), 260–267.
Cruden, D. M., & Varnes, D. J. (1996). Landslide types and processes. In A. K. Turner & R. L. Schuster (Eds.), Landslides: Investigation and mitigation (Transportation Research Board Special Report 247, pp. 36–75). National Academy Press.
Cutter, S. L., Boruff, B. J., & Shirley, W. L. (2003). Social vulnerability to environmental hazards. *Social Science Quarterly*, *84*(2), 242–261.
Cutter, S. L., & Finch, C. (2008). Temporal and spatial changes in social vulnerability. *Proceedings of the National Academy of Sciences*, *105*(7), 2301–2306.
Davis, M. (2006). *Planet of slums*. Verso.
Duncan, J. M., & Wright, S. G. (2005). *Soil strength and slope stability*. Wiley.
Fan, X., Scaringi, G., Korup, O., West, A. J., van Westen, C. J., Tanyas, H., Hovius, N., Hales, T. C., Jibson, R. W., Allstadt, K. E., Zhang, L., Evans, S. G., Xu, C., Li, G., Pei, X., Xu, Q., & Huang, R. (2019). Earthquake-induced chains of geologic hazards: Patterns, mechanisms, and impacts. Reviews of Geophysics, 57(2), 421–503. https://doi.org/10.1029/2018RG000626
Gesch, D., Oimoen, M., Greenlee, S., Nelson, C., Steuck, M., & Tyler, D. (2002). The National Elevation Dataset. *Photogrammetric Engineering & Remote Sensing*, *68*(1), 5–11.
Gill, J. C., & Malamud, B. D. (2014). Reviewing and visualizing the interactions of natural hazards. *Reviews of Geophysics*, *52*(4), 680–722.
Glade, T. (2003). Landslide occurrence as a response to land use change. *Geomorphology*, *56*(1–2), 77–89.
Greenway, D. R. (1987). Vegetation and slope stability. In M. G. Anderson & K. S. Richards (Eds.), *Slope stability*. Wiley.
Guardian. (2025, June 1). "This is ground zero": The Swiss village engulfed by a mountain. *The Guardian*.
Guzzetti, F., Carrara, A., Cardinali, M., & Reichenbach, P. (1999). Landslide hazard evaluation. *Geomorphology*, 31(1–4), 181–216.
Guzzetti, F., Reichenbach, P., Cardinali, M., & Galli, M. (2005). Probabilistic landslide hazard assessment at the basin scale. Earth Surface Processes and Landforms, *30*(4), 427–455.
Handwerger, A. L., Roering, J. J., & Schmidt, D. A. (2019). Kinematics of slow-moving landslides. *Nature Geoscience*, *12*, 555–560.
Hansen, M. C., Potapov, P. V., Moore, R., Hancher, M., Turubanova, S. A., Tyukavina, A., Thau, D., Stehman, S. V., Goetz, S. J., Loveland, T. R., Kommareddy, A., Egorov, A., Chini, L., Justice, C. O., & Townshend, J. R. G. (2013). High-resolution global maps of 21st-century forest cover change. Science, *342*(6160), 850–853.
Hardoy, J., & Satterthwaite, D. (2014). *Environmental problems of cities in the global south*. Routledge.

Hong, Y., Adler, R. F., & Huffman, G. J. (2006). Evaluation of satellite precipitation estimation. *Journal of Hydrometeorology*, *7*(6), 1238–1256.

Intrieri, E., Gigli, G., Mugnai, F., Fanti, R., & Casagli, N. (2013). Design and implementation of a landslide early warning system. Landslides, *10*(2), 143–155.

IPCC. (2022). *AR6 working group II: Impacts, adaptation and vulnerability*. Cambridge University Press.

Iverson, R. M. (2000). Landslide triggering by rainfall infiltration. *Water Resources Research*, *36*(7), 1897–1910.

Keefer, D. K. (1984). Landslides caused by earthquakes. *Geological Society of America Bulletin*, *95*(4), 406–421.

Keefer, D. K. (2002). Investigating landslides caused by earthquakes. *Engineering Geology*, *64*(2–3), 145–159.

Larsen, M. C., & Torres-Sánchez, A. J. (1998). The frequency and distribution of recent landslides in three montane tropical regions of Puerto Rico. Geomorphology, *24*(4), 309–331. https://doi.org/10.1016/S0169-555X(98)00023-3

Marais, A. (2025). *From risk to resilience: How community-led DRR drives formalization in informal settlements*. Unpublished independent study, Elliott School of International Affairs, The George Washington University.

Mejía, A. (2006). Integrated landslide risk management in Manizales, Colombia: The "Guardianas de la Ladera" program. *Proceedings of the International Conference on Disaster Reduction* (or related risk-management conference papers circulated through Manizales municipal programs).

Mejía Prieto, B., Giraldo Valencia, G. I., & Trujillo Gálvez, L. M. (2006). *Guardians of the slope*. Corporación para el Desarrollo de Caldas.

Montgomery, D. R., & Dietrich, W. E. (1994). A physically based model for shallow landsliding. *Water Resources Research*, 30(4), 1153–1171.

Montgomery, D. R., Schmidt, K. M., Dietrich, W. E., & McKean, J. (2000). Analysis of erosion and sediment yield. *Water Resources Research*, *36*(7), 1911–1926.

Moore, I. D., Grayson, R. B., & Ladson, A. R. (1991). Digital terrain modelling. *Hydrological Processes*, *5*(1), 3–30.

Nicholls, R. J., Wong, P. P., Burkett, V. R., Codignotto, J. O., Hay, J. E., McLean, R. F., Ragoonaden, S., & Woodroffe, C. D. (2007). Coastal systems and low-lying areas. In M. L. Parry, O. F. Canziani, J. P. Palutikof, P. J. van der Linden, & C. E. Hanson (Eds.), Climate change 2007: Impacts, *adaptation* and *vulnerability. Contribution of working group II to the fourth assessment report of the intergovernmental panel on climate change* (pp. 315–356). Cambridge University Press.

Pelletier, J. D., Barron-Gafford, G. A., Breshears, D. D., Brooks, P. D., Chorover, J., Durcik, M., & Troch, P. A. (2015). Landscape response to land-use change: Implications for soil redistribution and soil organic carbon across land-use transitions. *Earth Surface Processes and Landforms*, *40*(8), 1093–1105.

Petley, D. (2012). Global patterns of loss of life from landslides. *Geology*, *40*(10), 927–930.

Reichenbach, P., Rossi, M., Malamud, B. D., Mihir, M., & Guzzetti, F. (2018). A review of statistically based landslide susceptibility models. Earth-Science Reviews, *180*, 60–91.

Restrepo, C., Walker, L. R., Shiels, A. B., Bussmann, R., Claessens, L., Fisch, S., Lozano, P., Negi, G., Paolini, L., Poveda, G., Ramos-Scharrón, C., Richter, M., & Velázquez, E. (2009). Landslides and their ecological consequences. Journal of Ecology, *97*(3), 364–372.

Sidle, R. C., & Ochiai, H. (2006). *Landslides: Processes, prediction, and land use*. American Geophysical Union.

Swiss Federal Office for the Environment (FOEN). (2025). Avalanche risk monitoring and evacuation protocols. Bern, Switzerland: Swiss Federal Office for the Environment.

Tanyaş, H., Allstadt, K. E., & van Westen, C. J. (2017). An updated earthquake-induced landslide inventory. *Earth Surface Processes and Landforms*, *42*(13), 1996–2012.

Thomas, M. F. (1994). Geomorphology in the tropics: A study of weathering and denudation in low latitudes. John Wiley & Sons.

Turner, B. L., Lambin, E. F., & Reenberg, A. (2003). The emergence of land change science. *PNAS*, *100*(14), 8074–8079.

UNDRR. (2022). *Global assessment report on disaster risk reduction*. United Nations.

United Nations Human Settlements Programme (UN-Habitat). (2018). Addressing urban informality. UN-Habitat.

Vale, L. J., & Campanella, T. J. (Eds.). (2005). The resilient city: How modern cities recover from disaster. Oxford University Press.

Van Westen, C. J., Castellanos, E., & Kuriakose, S. L. (2008). Spatial data for landslide susceptibility, hazard, and vulnerability assessment: An overview. *Engineering Geology*, *102*(3–4), 112–131. https://doi.org/10.1016/j.enggeo.2008.03.010

Velásquez Barrero, L. S. (1998). Agenda 21 in Manizales. *Environment and Urbanization*, *10*(2), 9–36.

Wesely, J. (2021). Enabling environments for good practices in disaster risk management. *Environment and Urbanization*, *33*(2), 447–464.

Wong, P. P., Losada, I. J., Gattuso, J.-P., Hinkel, J., Khattabi, A., McInnes, K. L., Saito, Y., & Sallenger, A. H. (2014). Coastal systems and low-lying areas. In C. B. Field, V. R. Barros, D. J. Dokken, K. J. Mach, M. D. Mastrandrea, T. E. Bilir, M. Chatterjee, K. L. Ebi, Y. O. Estrada, R. C. Genova, B. Girma, E. S. Kissel, A. N. Levy, S. MacCracken, P. R. Mastrandrea, & L. L. White (Eds.), Climate change 2014: Impacts, *adaptation, and vulnerab*ility. Part A: *global and sectoral aspe*cts. Cambridge University Press.

World Bank. (2017). *Sierra Leone post-mudslide and floods needs assessment*. World Bank

Wu, T. H., McKinnell, W. P., & Swanston, D. N. (1979). Strength of tree roots and landslides on Prince of Wales Island, Alaska. Canadian Geotechnical Journal, *16*(1), 19–33. https://doi.org/10.1139/t79-003

10

Air Pollution and Traffic

Chapter Introduction

Air pollution is one of the most pervasive urban hazards of our time; often invisible, it is frequently overlooked as a source of pathology and too often normalized. In cities across the world, chronic exposure to air pollutants arises from a veritable crossroads of sources including vehicle emissions along congested roadways, industrial zones embedded within residential areas, and widespread reliance on informal or combustion-based energy systems for cooking, heating, and power generation. Unlike acute disasters, poor air quality rarely announces itself as a singular event, but instead accumulates quietly in lungs and bloodstreams, shaping long-term health outcomes and shortening lives in ways that are difficult to trace back to any single cause.

The irony is that many of the same networks that enable urban life also generate these risks. Roads, ports, rail yards, markets, and industrial corridors are essential to trade and economic growth, yet they simultaneously function as conduits for pollution and sometimes lethal exposure. These infrastructures command where emissions are produced, where they disperse, and which populations are hit the hardest. This is where geospatial analysis becomes indispensable. GIS allows researchers and planners to visualize the spatial logic of air pollution, linking emissions to land use and population vulnerability, and identifying intervention points where policies such as incentives or regulations can reduce harm.

10.1 Setting the Scene: Traffic, Air, and Everyday Exposure

On a summer afternoon in Los Angeles, traffic slows to a crawl along the 405. Heat radiates from the asphalt, engines idle, and a brownish haze settles over the basin. From a distance, the city appears functional, with cars moving, goods circulating, commuters inching home. Yet the air tells a different story. Nitrogen dioxide levels spike near freeways, carbon monoxide accumulates in traffic corridors, and fine particulates linger in neighborhoods downwind

DOI: 10.1201/9781032636948-10

of congested roads. For millions of residents, this is not a temporary emergency but a routine exposure, repeated day after day.

Los Angeles has long served as a bellwether for the relationship between urban mobility and air pollution. Decades of monitoring show that while regulatory efforts have reduced some pollutants since the 1970s, traffic-related emissions remain a dominant contributor to localized exposure, particularly near major roadways (Bureau of Transportation Statistics [BTS], 2023). Studies using roadside monitors and satellite-derived observations demonstrate sharp spatial gradients: concentrations of NO_2 and CO fall rapidly with distance from highways, creating a patchwork of exposure tied closely to where people live, work, and travel (Fecht et al., 2019). Figure 10.1 depicts the effects of various air pollutant gases including fine particulates, ground-level ozone, and carbon monoxide on the human body.

This spatial unevenness matters. Traffic-related air pollutants (TRAPs), which are primarily composed of nitrogen dioxide and carbon monoxide alongside ultrafine particles, are strongly associated with respiratory illness, cardiovascular disease, and adverse birth outcomes (Khreis et al., 2020). Unlike regional smog episodes, TRAP exposure is often chronic and highly localized, affecting residents of corridor-adjacent neighborhoods, as well as transit workers, street vendors, and schoolchildren whose classrooms are often located near traffic-clogged arterial roads. The hazard is not just the volume of pollution, but its persistence and predictability.

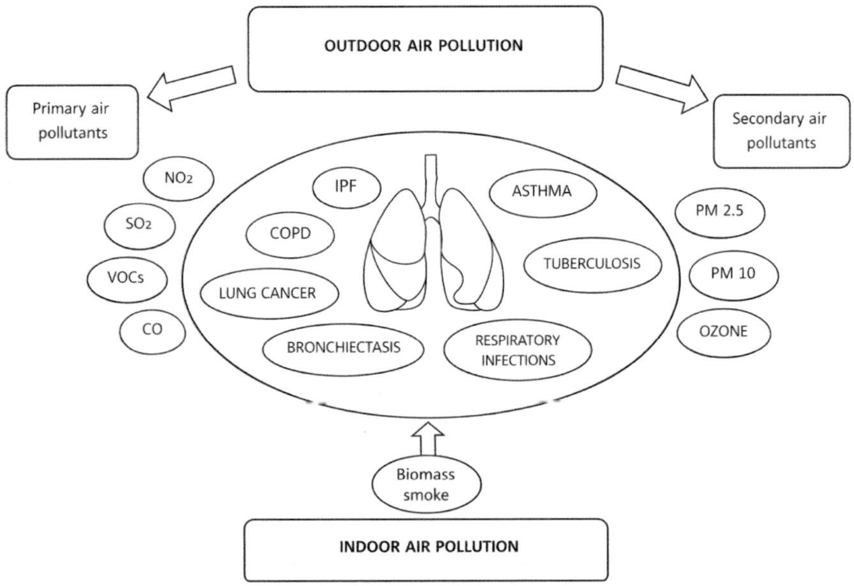

FIGURE 10.1
Air pollution exposure chart. (Redrawn from Bălă et al., 2021.)

Recent years have added a new layer of complexity. Wildfire smoke, driven by climate change, has increasingly blanketed Southern California, producing prolonged "code red" air quality days that compound existing traffic emissions (WHO, 2022). During these episodes, particulate matter concentrations can exceed health-based thresholds for days or weeks at a time, overwhelming regulatory gains made through vehicle standards alone. Health systems report spikes in emergency room visits for asthma and cardiovascular distress, particularly among older adults and those with pre-existing conditions (WHO, 2022).

What makes Los Angeles instructive for cities globally is not simply the scale of its traffic, but the clarity of its data. Dense monitoring networks, long-term trend analysis, and high-resolution satellite products make it possible to visualize how roads become exposure corridors and how mobility systems shape public health risk (Fecht et al., 2019). In many rapidly growing cities elsewhere, similar processes are underway. But without comparable monitoring capacity, the hazard remains less visible and therefore less actionable.

Air pollution, then, is not an abstract environmental problem. It is produced by the everyday functioning of cities and experienced unevenly across urban space. The Los Angeles case underscores a central theme of this chapter: traffic and air quality are inseparable, and GIS provides the tools to make that relationship legible, measurable, and ultimately actionable.

10.2 Unpacking the Threat: How Air Pollution Affects Cities

Urban air pollution is not produced by a single source or process. Rather, it emerges from the interaction of multiple emission sources, atmospheric conditions, and urban spatial arrangements. In cities in the Global South, this interaction is often intensified by rapid growth, aging infrastructure, and regulatory gaps, resulting in exposure environments that differ fundamentally from those in wealthier cities. Understanding air pollution as a *spatially structured threat*—rather than a uniform environmental condition—is essential for effective analysis and intervention (WHO, 2022).

10.2.1 Sources and Processes of Air Pollution

Urban air pollution arises from multiple, interacting emission sources whose impacts are shaped by atmospheric processes and urban form. Vehicle exhaust remains the most pervasive source in most cities, producing carbon monoxide (CO), nitrogen oxides (NO_x), volatile organic compounds (VOCs), and fine particulates. Emissions intensify under congestion, frequent stopping and starting, poor vehicle maintenance, and the prevalence of older or diesel-powered fleets—conditions common in rapidly growing cities with limited inspection regimes (Khreis et al., 2020; WHO, 2022).

Industrial activity introduces a second major source category. Power plants, refineries, cement works, ports, slaughterhouses, and manufacturing zones emit sulfur dioxide (SO_2), nitrogen oxides, particulates, and process-specific pollutants. In many Global South cities, industrial land uses are interwoven with residential neighborhoods, producing sustained exposure rather than isolated hotspots. Weak zoning enforcement and informal industrial operations further blur the separation between production and habitation (WHO, 2022).

Domestic energy use constitutes a third, often underestimated source. The widespread use of biomass, charcoal, kerosene, and low-quality fuels for cooking and heating produces fine particulates and carbon monoxide at the household level. As depicted in Figure 10.2, cooking over an indoor wood fire is a common practice in many parts of the world, including in rural India. Indoor emissions frequently spill into outdoor environments, particularly in dense settlements, contributing to neighborhood-scale pollution burdens that disproportionately affect women, children, and older adults (WHO, 2022).

FIGURE 10.2
Cookstoves as domestic source of emissions. (Photo depicts cooking on a clay stove with firewood in a village in north Andhra. Creative Commons CC-BY-4.0. Author: Saiphani02.)

TABLE 10.1

Major Urban Air Pollutants: Sources and Impacts

Pollutant	Primary Urban Sources	Key Health Impacts	Spatial Characteristics
SO_2	Power plants, refineries, diesel fuel, industry	Respiratory irritation, asthma exacerbation	Concentrated near industrial zones
NO_x (NO, NO_2)	Vehicle exhaust, power generation	Asthma, reduced lung function, precursor to ozone	Strong roadside gradients
O_3 (ozone)	Secondary pollutant formed from NO_x+VOCs	Lung inflammation, reduced lung capacity	Regional, diurnal patterns
PM_{10}	Dust, construction, unpaved roads	Respiratory disease	Local to neighborhood scale
$PM_{2.5}$	Combustion (vehicles, industry, biomass)	Cardiovascular disease, premature mortality	Local to regional
CO	Incomplete combustion (traffic, cooking)	Cardiovascular stress, neurological effects	Highly localized near sources

Source: WHO (2022); Khreis et al. (2020).

Finally, construction activity, mining, and unpaved roads generate coarse and fine dust, especially during dry seasons. These sources may appear episodic, but in rapidly expanding cities, they become quasi-permanent contributors to particulate exposure. Once emitted, pollutants disperse unevenly depending on wind, temperature, topography, and the built environment. Temperature inversions, urban canyons, and coastal breeze cycles can trap or redistribute pollutants, producing sharp spatial and temporal contrasts in exposure (Khreis et al., 2020).

Table 10.1 lists major air pollution threats in cities.

At its most basic level, urban air pollution originates from four overlapping source categories: vehicle emissions, industrial activity, domestic energy use, and construction or land disturbance. Motor vehicles emit carbon monoxide (CO), nitrogen oxides (NO_x), and VOCs, particularly under conditions of congestion and idling. Industrial facilities release sulfur dioxide (SO_2), particulates, and process-specific pollutants, often from fixed locations that shape persistent exposure zones. Domestic combustion—especially biomass and charcoal used for cooking and heating—introduces fine particulates and carbon monoxide directly into living spaces. Construction activity and unpaved roads add coarse and fine dust, particularly during dry seasons (WHO, 2022).

These emissions do not remain fixed at their points of origin. Once released, pollutants disperse according to wind speed and direction, temperature, topography, and the built form of the city. Temperature inversions can trap pollutants near the ground, while dense building clusters create "urban canyons" that inhibit ventilation and concentrate emissions along streets and corridors. Coastal cities experience land-sea breeze cycles that redistribute pollution diurnally, while inland cities may accumulate pollutants during calm atmospheric conditions. As a result, exposure varies not only across neighborhoods but also across hours of the day and seasons of the year (Khreis et al., 2020).

Air pollution is also chemically dynamic. Primary pollutants such as CO, NO_2, and SO_2 are emitted directly from sources, while secondary pollutants, including ozone (O_3) and some fine particulates, form through chemical reactions in the atmosphere. This distinction matters for spatial analysis: primary pollutants tend to be highly localized and closely tied to specific sources, whereas secondary pollutants may exhibit broader regional patterns. Cities experiencing both heavy traffic and strong sunlight, such as those in tropical and subtropical regions, are particularly prone to elevated ozone formation (WHO, 2022).

Critically, exposure is shaped as much by urban form and social conditions as by emissions themselves. Populations living near major roads, industrial estates, ports, and markets often experience higher pollutant concentrations while simultaneously facing barriers to healthcare, political influence, and mitigation measures. In this sense, air pollution is not only an environmental hazard but also a spatial expression of inequality. GIS-based analysis makes this visible by linking pollutant concentrations to specific land uses, and especially transportation systems, laying the groundwork for targeted, evidence-based action.

TRAFFIC-RELATED AIR POLLUTANTS (TRAPs)

TRAPs refer to a suite of pollutants emitted directly from motor vehicles and concentrated along roadways and transport corridors. Unlike regional air pollution, which may be shaped by atmospheric chemistry and long-range transport, TRAPs are highly localized and strongly patterned by traffic volume, vehicle type, speed, and congestion. The most commonly cited TRAPs include carbon monoxide (CO) and nitrogen dioxide (NO_2), along with ultrafine particles and other combustion byproducts associated with gasoline and diesel engines (Khreis et al., 2020).

What distinguishes TRAPs from broader air quality metrics is their spatial intensity and persistence. Concentrations of CO and NO_2 are typically highest within a few hundred meters of major roads and decline sharply with distance, creating narrow exposure corridors that intersect unevenly with residential areas, schools, markets, and informal workplaces. Epidemiological studies link TRAP exposure to asthma incidence, impaired lung development in children, cardiovascular disease, and adverse birth outcomes, even in cities that meet regional air quality standards (Khreis et al., 2020, Anenberg et al., 2019, Goldberg et al. 2021).

From a geospatial perspective, TRAPs are especially well suited to GIS-based analysis. Road networks, traffic counts, vehicle fleet composition, and land-use data can be combined to model exposure surfaces that more accurately reflect everyday mobility patterns than citywide averages. In many Global South cities—where congestion is chronic, vehicle regulation uneven, and monitoring sparse—TRAP-focused mapping provides a practical entry point for identifying high-risk populations and evaluating interventions such as traffic rerouting, low-emission zones, or pedestrianization (Khreis et al., 2020).

WHAT $PM_{2.5}$ MISSES

Fine particulate matter ($PM_{2.5}$) is often treated as the definitive indicator of urban air pollution risk. This emphasis is understandable, as $PM_{2.5}$ is strongly associated with cardiovascular and respiratory disease, and it is one of the few pollutants that can be estimated consistently at high spatial resolution using satellite data. As a result, $PM_{2.5}$ frequently anchors global comparisons and urban air quality indices.

However, a neighborhood-scale study of Richmond, Virginia illustrates the limits of this approach. In a senior capstone project, Malstead (2022) analyzed multiple pollutants and found that gases commonly associated with traffic and industrial activity, including nitric oxide (NO), carbon monoxide (CO), carbon dioxide (CO_2), nitrogen dioxide (NO_2), ozone (O_3), and sulfur dioxide (SO_2), often played a more significant role in shaping localized exposure patterns than $PM_{2.5}$ alone. Areas with elevated health risks were not always those with the highest particulate concentrations, but rather those situated near transportation corridors and industrial land uses where gaseous pollutants dominated.

For geospatial analysis, the implication is clear. Relying solely on PM_{25} can obscure important spatial dynamics of exposure, particularly in cities where traffic congestion, informal industry, or diesel fleets are major pollution sources. Multi-pollutant mapping—aligned with road networks, land use, and population vulnerability—provides a more accurate representation of how air pollution is experienced on the ground (Malstead, 2022).

10.2.2 Urban Snapshots: Chronic Exposure in Practice

10.2.2.1 Bangkok

Bangkok's air pollution problem reflects the interaction of traffic congestion, industrial activity, and meteorology. High vehicle density, diesel-powered buses and trucks, and frequent congestion generate persistent NO_2 and $PM_{2.5}$

hotspots along major corridors. Seasonal temperature inversions and low wind speeds exacerbate pollutant accumulation, while rapid urban expansion has increased commuting distances and exposure durations (WHO, 2022).

10.2.2.2 New Delhi

Delhi consistently ranks among the most polluted megacities in the world. Vehicular emissions interact with industrial sources, coal-fired power plants, construction dust, and seasonal agricultural burning in surrounding regions. Winter inversions trap pollutants near the surface, producing prolonged high-exposure episodes dominated by $PM_{2.5}$ and NO_x. Despite policy interventions, the sheer scale of traffic and energy demand continues to overwhelm mitigation efforts (WHO, 2022).

10.2.2.3 Beijing

Beijing's air quality challenges have historically been driven by coal combustion, heavy industry, and vehicle emissions. Aggressive regulatory measures have reduced some pollutants, yet traffic-related NO_2 and ozone remain persistent concerns. The city illustrates how secondary pollutants such as ozone can increase even as primary particulate emissions decline, highlighting the complexity of atmospheric chemistry in large urban regions (WHO, 2022).

10.2.3 Vehicles and Colonial Urban Form in the Global South

The spatial relationship between traffic and air pollution in the Global South cannot be understood without reference to the colonial origins of many contemporary cities. Large numbers of African, Asian, and Latin American cities were designed or reconfigured during the late colonial period, often before mass motorization but at the dawn of the internal combustion age. These cities were planned to serve administrative control, resource extraction, and limited settler populations, not the millions of residents who now call them home (WHO, 2022).

Colonial planning regimes prioritized separation: cordon sanitaires, racial zoning, and administrative districts physically distanced colonial elites from indigenous populations. Roads and rail lines were laid out to connect ports, government centers, and the extractive hinterlands, embedding mobility corridors that continue to structure urban life. As cities expanded after independence, these corridors became backbones of metropolitan transport systems, increasingly dominated by cars, minibuses, and trucks (WHO, 2022).

The rapid uptake of motor vehicles in the postwar period transformed these inherited layouts. Roads claimed disproportionate amounts of urban land, often at the expense of housing, public space, and non-motorized transport. Central business districts, markets, ports, and industrial estates became focal points for traffic convergence, intensifying emissions in precisely the areas where economic activity and population density overlapped. In countless cities across Africa and South Asia, truck routes and wholesale distribution systems pass directly through dense neighborhoods, producing sustained exposure to TRAPs (Khreis et al., 2020).

Unlike many cities in the Global North, where car dependence followed suburbanization and rising incomes, motorization in the Global South often preceded the development of comprehensive transit systems. Informal minibuses, motorcycles, and aging vehicle fleets filled mobility gaps, providing essential access while also generating high per-vehicle emissions. Regulatory capacity frequently lagged behind growth, allowing older, higher-emitting vehicles to remain in circulation far longer than in wealthier contexts (WHO, 2022).

The result is a distinctly spatial form of environmental inequality. Exposure to traffic-related pollution is highest where mobility systems intersect with population density, especially as it is found in urban areas with low-income housing, informal employment, and limited political influence. Air pollution thus reflects not only how cities move but also how they were designed, for whom, and under what historical conditions. GIS-based analysis is uniquely positioned to trace these layered geographies, linking colonial legacies, contemporary traffic flows, and present-day health risks into a coherent spatial narrative.

10.2.4 Impacts of the Air Pollution Threat in Cities

The impacts of urban air pollution extend far beyond degraded environmental quality. In cities, polluted air functions as a slow-moving but persistent hazard, shaping health outcomes, economic productivity, and patterns of inequality over time. Unlike episodic disasters, air pollution exerts its effects cumulatively, with daily exposure compounding across weeks, years, and lifetimes. Urban form intensifies these impacts by concentrating emissions, populations, and vulnerable groups within limited geographic space (WHO, 2022).

Public health consequences are the most extensively documented. Exposure to fine particulate matter, nitrogen dioxide, ozone, and carbon monoxide is associated with elevated risks of asthma, chronic obstructive pulmonary disease, cardiovascular disease, stroke, and premature mortality. Children are particularly vulnerable due to developing respiratory systems, while older adults and individuals with pre-existing conditions face heightened risks of hospitalization and death during periods of poor air quality (WHO, 2022). These outcomes are not evenly distributed across the city. Residents living near major roads, industrial estates, ports, and logistics hubs experience consistently higher exposure levels, often without corresponding access to healthcare or mitigation resources (Khreis et al., 2020).

Air pollution also operates through economic and labor-related pathways. Reduced lung function and cardiovascular strain diminish worker productivity, particularly in occupations requiring prolonged outdoor exposure such as construction, transport, street vending, and informal services. In many cities in the Global South, a large share of the workforce encounters polluted air throughout the workday, magnifying health impacts while increasing vulnerability to lost income. At the urban scale, these individual effects aggregate into substantial productivity losses and rising healthcare expenditures, reinforcing cycles of poverty and inequality (WHO, 2022).

Educational outcomes represent another, often overlooked, dimension of impact. Schools located near traffic corridors or industrial zones expose children to elevated pollution levels during critical periods of cognitive development. Research links air pollution exposure to reduced academic performance, increased absenteeism, and long-term cognitive effects. Because school locations frequently mirror residential patterns, pollution-related educational impacts tend to fall disproportionately on lower-income neighborhoods, contributing to the intergenerational reproduction of spatial inequality (WHO, 2022).

Urban infrastructure systems are affected as well. Air pollution accelerates the deterioration of buildings, bridges, and cultural heritage sites through chemical corrosion and particulate deposition. In historic city centers, especially those with stone architecture, sulfur dioxide and ozone contribute to surface erosion and structural weakening. These processes impose additional maintenance costs on already resource-constrained municipal governments while threatening irreplaceable cultural assets (WHO, 2022).

Psychological and social impacts, though less easily quantified, are increasingly recognized. Persistent exposure to poor air quality contributes to stress, anxiety, and diminished well-being, particularly during prolonged haze events or wildfire smoke episodes. Reduced visibility, odor, and physical discomfort alter how residents perceive and use urban space, discouraging outdoor activity and social interaction. In this way, air pollution reshapes not only health outcomes but also the everyday rhythms of urban life (WHO, 2022).

Climate change acts as a force multiplier for many of these impacts. Rising temperatures accelerate ozone formation, while prolonged droughts increase dust and wildfire risk. In cities already burdened by traffic emissions and industrial pollution, climate-driven extremes can push air quality beyond critical thresholds for extended periods. These compound events strain health systems and expose the limits of reactive, short-term mitigation strategies (Khreis et al., 2020; WHO, 2022).

From a spatial perspective, the defining feature of air pollution impacts in cities is their uneven distribution. Exposure surfaces align closely with transportation networks, land-use patterns, and socio-economic gradients. GIS-based analyses consistently show that the greatest health burdens fall on populations with the least capacity to avoid or mitigate exposure. This spatial injustice challenges the notion of air pollution as a diffuse or universal problem and instead frames it as a deeply place-based urban threat (Khreis et al., 2020).

Taken together, these impacts underscore why air pollution must be understood as an urban systems problem rather than a narrow environmental issue. Its effects ripple through health, labor, education, infrastructure, and social life, reinforcing existing inequalities and shaping long-term urban trajectories. Mapping these impacts—linking emissions to exposure, exposure to vulnerability, and vulnerability to outcomes—is therefore not an academic exercise but a prerequisite for effective and equitable urban action.

AN AFFECTED PLACE—NEW DELHI, INDIA

New Delhi illustrates how traffic, industry, domestic energy use, and atmospheric conditions can converge to produce extreme and persistent urban air pollution. Figure 10.3 shows an activity-filled scene from Old Delhi. For much of the year, residents experience pollutant concentrations far exceeding health-based guidelines, making air quality

FIGURE 10.3
New Delhi traffic and air pollution photo. (Old Delhi photo, 3032×1949. CC-BY-2.0. Photo by Ruffin_Ready.)

a routine concern rather than an episodic crisis. The city's exposure profile is shaped by a combination of dense traffic, coal-fired power generation, construction activity, informal industry, and seasonal agricultural burning in surrounding regions (WHO, 2022).

Traffic plays a central role in Delhi's pollution burden. Rapid motorization, high congestion levels, and a large fleet of diesel-powered trucks, buses, and older vehicles generate substantial emissions of nitrogen dioxide (NO_2), carbon monoxide (CO), and fine particulate matter. Major arterial roads and ring highways function as continuous emission corridors, with pollutant concentrations declining sharply with distance but affecting millions who live, work, or attend school nearby. These TRAPs interact with industrial emissions and dust to create complex exposure surfaces across the metropolitan region (Khreis et al., 2020).

Seasonal atmospheric conditions amplify these emissions. During winter months, temperature inversions trap pollutants close to the ground, while low wind speeds limit dispersion. At the same time, post-harvest agricultural burning in neighboring states introduces additional particulates that are transported into the city. The result is prolonged episodes of extreme pollution dominated by $PM_{2.5}$ and NO_x, often lasting weeks rather than days (WHO, 2022). These episodes regularly push air quality indices into "severe" categories, prompting school closures and emergency advisories.

The health impacts are profound and unevenly distributed. Children in Delhi show elevated rates of asthma and impaired lung development, while adults face increased risks of cardiovascular disease and stroke. Outdoor workers—including traffic police, street vendors, and construction laborers—experience especially high cumulative exposure due to prolonged time spent near emission sources. Lower-income neighborhoods, frequently located near major roads or industrial zones, bear a disproportionate share of this burden while having limited access to healthcare or mitigation options such as air filtration (WHO, 2022).

Delhi has implemented a range of policy responses, including vehicle emission standards, fuel quality improvements, restrictions on older vehicles, and periodic traffic rationing schemes. While these measures have produced incremental improvements, their effectiveness is constrained by the scale of emissions and the structural dependence on road-based transport. The city's experience underscores a key lesson for urban air quality management: without addressing the spatial organization of mobility, land use, and energy systems, regulatory gains alone are unlikely to produce sustained improvements.

From a geospatial perspective, New Delhi demonstrates why air pollution must be analyzed as a spatially embedded urban threat.

Emissions, exposure, and vulnerability align closely with transportation networks, land-use patterns, and socio-economic gradients. GIS-based mapping of these relationships has become central to diagnosing the problem, communicating risk, and evaluating potential interventions—making Delhi both a cautionary example and a critical laboratory for urban air pollution research (Khreis et al., 2020; WHO, 2022).

10.2.5 Increasing Sub-Threat: Wildfire Smoke as an Urban Air Pollution Multiplier

Wildfire smoke has emerged as a rapidly intensifying sub-threat to urban air quality, transforming what were once episodic rural hazards into recurrent urban exposure events. Climate change has increased the frequency, intensity, and duration of wildfires across many regions, while expanding urban footprints place more people directly in harm's way. For cities, wildfire smoke represents not a separate hazard but a powerful amplifier of existing air pollution problems—particularly those driven by traffic and industrial emissions (WHO, 2022).

Wildfire smoke is dominated by fine particulate matter ($PM_{2.5}$), along with carbon monoxide and a complex mixture of organic compounds. Unlike many urban pollutants, smoke plumes can travel hundreds or even thousands of kilometers, affecting cities far removed from the fire source. Once present, smoke interacts with local emissions, pushing already elevated pollutant concentrations beyond critical health thresholds for extended periods. In traffic-heavy cities, this creates compound exposure environments where background pollution and episodic smoke reinforce one another (WHO, 2022).

Urban populations are particularly vulnerable to these compound events. During wildfire smoke episodes, emergency room visits for asthma, cardiovascular distress, and respiratory infections increase sharply, even among individuals without prior diagnoses. The health burden is greatest for children, older adults, outdoor workers, and those with limited ability to modify behavior or access indoor air filtration. In many Global South cities, housing quality and energy insecurity make it difficult to seal indoor environments against smoke intrusion, further increasing exposure (WHO, 2022).

From a spatial perspective, wildfire smoke challenges conventional urban air quality management. Unlike TRAPs, smoke does not correlate with municipal boundaries or zoning regimes. Its impacts vary with wind patterns, topography, and urban form, producing highly uneven exposure surfaces that shift daily or even hourly. GIS-based analysis combining satellite fire detection, aerosol optical depth, wind fields, and population data has become essential for tracking smoke movement and identifying high-risk neighborhoods in real time (WHO, 2022).

As wildfire risk continues to grow, cities that already struggle with chronic air pollution face a future of more frequent compound exposure events. This sub-threat underscores the need to integrate climate adaptation into urban air quality planning and to treat air pollution not as a static condition, but as a dynamic hazard shaped by interacting environmental systems.

10.2.6 Discussion: Air Pollution as a Spatial Expression of Inequality

Taken together, the preceding sections demonstrate that urban air pollution is not a singular phenomenon but a layered and spatially differentiated threat. Emissions arise from multiple sources including traffic, industry, domestic energy use, and land disturbance, yet their impacts depend on how cities are organized and for whom they function. Roads, ports, markets, and industrial corridors concentrate pollutants in specific locations, while atmospheric processes and urban form shape their dispersion. The result is a geography of exposure that mirrors broader patterns of inequality (Khreis et al., 2020; WHO, 2022).

One of the central insights from urban air pollution research is that vulnerability is place-based. Residents living near major transport corridors or industrial zones are exposed to higher pollutant concentrations regardless of citywide averages. These same areas often coincide with lower-income housing, informal settlements, and limited access to healthcare or political influence. Air pollution thus becomes both a cause and a consequence of urban deprivation, reinforcing existing social and economic divides (WHO, 2022).

The tendency to rely on single metrics, such as $PM_{2.5}$ or citywide air quality indices, can obscure these spatial dynamics. As the Richmond and TRAPs examples illustrate, localized exposure to gaseous pollutants and ultrafine particles may drive health impacts even when particulate concentrations appear moderate. A geospatial perspective reveals these mismatches by linking pollutant data to land use, traffic flows, and population characteristics, moving analysis closer to lived experience (Khreis et al., 2020).

Climate change further complicates this picture. Rising temperatures accelerate ozone formation, while wildfires introduce episodic yet severe pollution events that compound chronic exposure. These interactions challenge traditional regulatory approaches that assume relatively stable background conditions. For cities in the Global South, where monitoring capacity is often limited, the risk is that compound hazards remain under-detected until health impacts become unavoidable (WHO, 2022).

Ultimately, air pollution forces a reconsideration of how cities balance mobility, growth, and public health. The problem is not simply that vehicles and industries pollute, but that urban systems are designed in ways that prioritize movement and throughput over exposure reduction. Addressing air pollution therefore requires interventions that operate at multiple scales—from vehicle standards and fuel quality to land-use planning and transport design. GIS provides the analytical bridge between these scales, enabling cities to diagnose problems, visualize trade-offs, and evaluate the spatial consequences of policy choices.

10.3 Modeling Threat Risk in Geospatial Terms

Understanding air pollution as a spatially structured urban threat raises a critical question: how can cities systematically identify where risks are greatest and where interventions will have the most impact? Answering this requires moving beyond descriptive accounts of pollution sources and impacts toward analytical frameworks that integrate emissions, exposure, and vulnerability.

Geospatial modeling provides this framework. By combining traffic networks, industrial land use, atmospheric data, and population characteristics, GIS enables analysts to map exposure surfaces that reflect how pollution is actually experienced across urban space. These models make it possible to identify hotspots, evaluate cumulative exposure along daily travel routes, and test scenarios such as traffic rerouting, low-emission zones, or industrial relocation.

In the following section, the chapter turns from diagnosis to analysis. It introduces conceptual tools for modeling air pollution risk, then it outlines some project design questions and presents a practical workflow for building air pollution maps using open data. The goal is not only to measure air pollution but also to translate complex environmental data into actionable spatial insight.

10.3.1 Conceptual Tools for Modeling Air Pollution Risk

Modeling urban air pollution requires conceptual tools that link emissions, movement, and exposure across space and time. Key concepts include source–receptor relationships, which describe how pollutants travel from emission points to exposed populations, and exposure surfaces, which integrate pollutant concentration with population density and activity patterns. Network-based exposure treats roads and commuting routes as dynamic risk corridors rather than static lines. Cumulative exposure emphasizes repeated, low-level contact over time rather than episodic peaks. Together, these concepts shift analysis from citywide averages toward place-based risk, aligning air pollution modeling with lived urban experience (Khreis et al., 2020; WHO, 2022).

10.3.2 Project Design Questions

Effective geospatial analysis of air pollution begins with clear project design questions that define scope, scale, and purpose. Analysts must first ask which pollutants matter most in the local context—particulates, gaseous pollutants, or TRAPs —and whether the focus is chronic exposure or episodic events such as wildfire smoke. Spatial questions follow: Where are emissions generated, and how do traffic networks, land use, and atmospheric conditions shape dispersion? Who is exposed, when, and for how long? Temporal resolution is equally important, as daily commuting patterns and seasonal inversions can alter exposure significantly. Finally, projects should ask how results will be used, whether to identify hotspots, evaluate policy scenarios, or prioritize interventions, ensuring that models remain actionable rather than purely descriptive (Khreis et al., 2020; WHO, 2022).

10.3.3 Lab: Building an Air Pollution Map with Open Data

This lab introduces a practical workflow for mapping urban air pollution using openly available geospatial data. The objective is to move from abstract discussion of emissions and exposure to a concrete, spatial representation of air pollution risk that can inform analysis and decision-making. While the example can be adapted to many cities, the methods are designed to be transferable to data-scarce contexts common in the Global South.

The first step is defining the study area and pollutant of interest. Analysts should select a city or metropolitan region and identify one or more pollutants relevant to local conditions, such as $PM_{2.5}$ for combustion-related exposure or nitrogen dioxide (NO_2) for traffic-related pollution. Satellite-derived air quality products provide a consistent starting point. The Sentinel-5P TROPOMI instrument offers daily global coverage for NO_2, CO, and ozone, while $PM_{2.5}$ surfaces can be derived from aerosol optical depth products combined with ground observations where available (WHO, 2022).

Next, assemble core spatial layers that contextualize pollution patterns. Road networks can be sourced from open datasets such as OpenStreetMap or the Overture Maps Foundation, which provide information on road hierarchy and connectivity. Population density layers, available from sources such as WorldPop or national statistical organizations (NSOs), allow analysts to estimate potential exposure rather than pollution alone. Land-use data help distinguish residential, industrial, and commercial zones, clarifying where emissions and people intersect (WHO, 2022).

With these layers in place, analysts can begin exploratory spatial analysis. One common approach is hotspot mapping, which overlays pollutant concentration surfaces with population density to identify neighborhoods facing elevated exposure. Buffer analysis around major roads, typically ranging from 100 to 500 m, helps visualize the spatial footprint of TRAPs. These buffers can be weighted by road type or estimated traffic volume to better approximate emission intensity (Khreis et al., 2020).

More advanced analyses incorporate mobility. Network-based exposure modeling treats roads as dynamic systems rather than static features. By estimating exposure along commuting routes or public transit corridors, analysts can assess how daily movement patterns shape cumulative risk. Even in the absence of detailed traffic counts, proxy measures such as road classification and connectivity can produce informative exposure surfaces in data-limited settings (Khreis et al., 2020).

Visualization is a critical final step. Effective maps emphasize gradients, uncertainty, and context rather than false precision. Analysts should avoid overinterpreting satellite-derived values at fine scales and instead use them to highlight relative patterns of risk. Clear legends, explanatory captions, and transparency about data limitations are essential, particularly when maps are intended to inform policy or public discussion (WHO, 2022).

Figures 10.4 and 10.5 depict air quality and traffic levels for Paris, compiled from open sources.

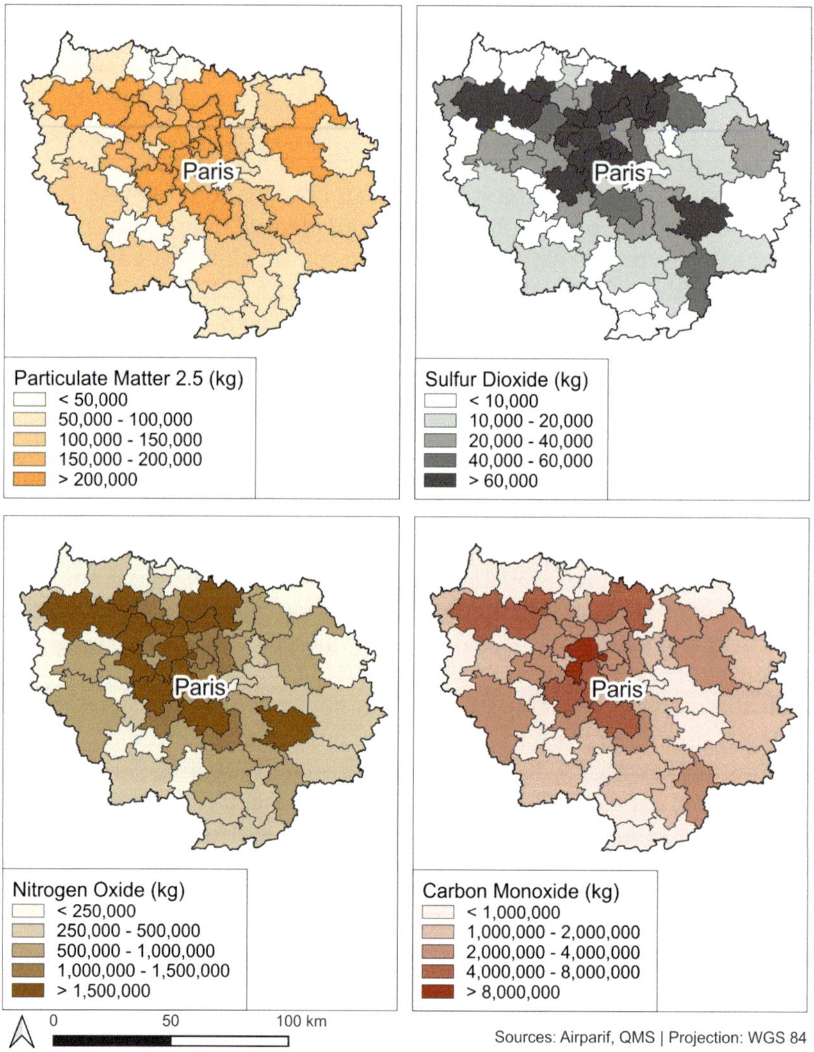

FIGURE 10.4
Paris AQI maps.

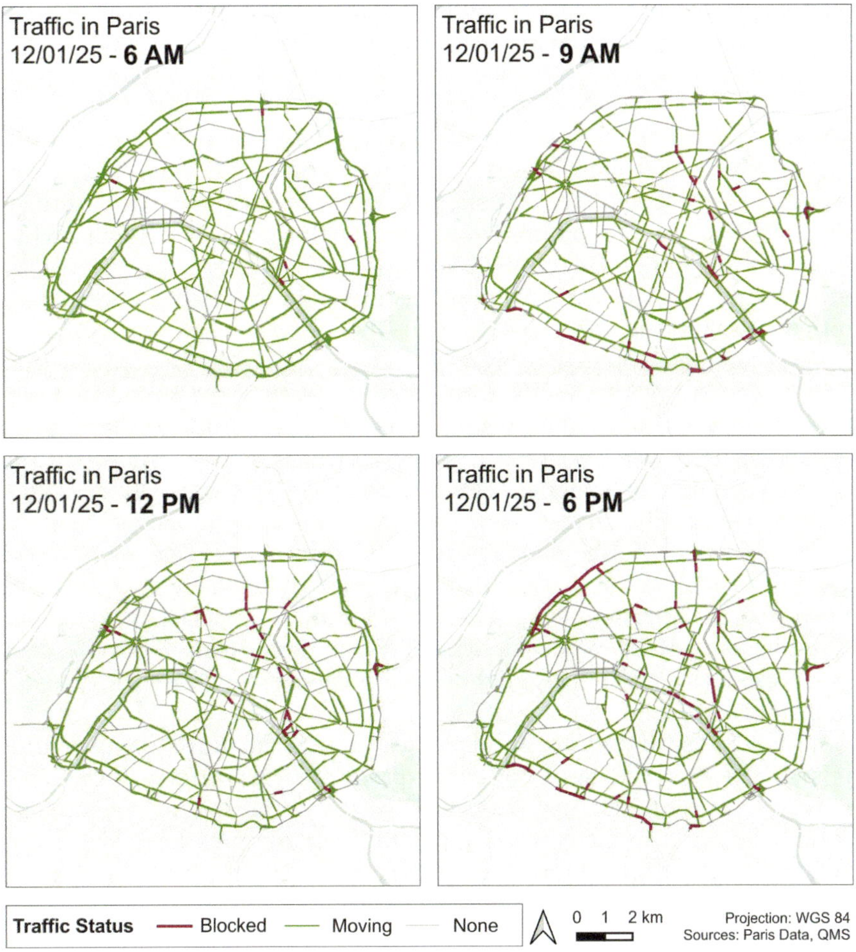

FIGURE 10.5
Paris traffic maps. (Both are by Maya O'Brien.)

This lab demonstrates how open data and GIS can make urban air pollution visible, measurable, and spatially explicit. While such maps do not replace regulatory monitoring or epidemiological studies, they provide a powerful entry point for diagnosing exposure patterns, prioritizing interventions, and linking air quality to broader questions of urban design and equity.

10.3.4 Lab Extension: Adapting the Air Pollution Mapping Workflow to a South Asian City

This lab extension applies the open-data workflow to a large Asian city, with New Delhi as an illustrative example. Delhi is particularly well suited to this exercise due to its chronic air pollution, strong traffic signal, and availability of complementary satellite and ground-based data. The goal is not to produce definitive exposure estimates, but to demonstrate how GIS can integrate diverse datasets to reveal spatial patterns of risk.

Begin by defining the study area as the National Capital Territory of Delhi or the broader metropolitan region. Select two pollutants that reflect different processes: $PM_{2.5}$ to capture combustion and regional haze, and nitrogen dioxide (NO_2) to represent traffic-related air pollution. Sentinel-5P provides daily NO_2 data at neighborhood-scale resolution, while $PM_{2.5}$ surfaces can be derived from global gridded products that combine satellite observations with atmospheric models (WHO, 2022).

Next, assemble contextual layers. Road networks can be sourced from OpenStreetMap, with classification used as a proxy for traffic intensity. Major arterials, ring roads, and highway corridors should be highlighted. Population density layers from WorldPop allow analysts to shift from pollution concentration to potential exposure. Land-use data help distinguish residential areas from industrial zones, logistics hubs, and commercial districts.

The initial analysis should focus on spatial coincidence. Overlay NO_2 concentrations with major road corridors to identify traffic-related exposure hotspots. Create buffers (e.g., 200–500 m) around high-capacity roads and examine how population density varies within these zones. For $PM_{2.5}$, compare seasonal averages to highlight winter inversion periods when exposure intensifies across the city (WHO, 2022).

An optional advanced step incorporates temporal analysis. Using monthly or seasonal averages, students can explore how exposure patterns shift between monsoon and dry seasons, illustrating the role of meteorology. Wind direction data can be used qualitatively to interpret downwind exposure zones, even without formal dispersion modeling.

The final deliverable should include at least two maps: one emphasizing traffic-related pollution (NO_2) and another showing broader particulate exposure ($PM_{2.5}$), accompanied by a short written interpretation. Students should be encouraged to reflect on data limitations, including the mismatch between satellite resolution and street-level exposure, and to consider how additional data—such as traffic counts or household fuel use—could refine the analysis.

This adapted lab demonstrates how open geospatial tools can make complex urban air pollution dynamics visible, even in megacities with limited monitoring infrastructure. It reinforces a central lesson of this chapter: air pollution risk is spatially patterned, policy-relevant, and well suited to GIS-based investigation.

10.3.5 Advanced Modeling Approaches

Basic air pollution maps reveal where pollutants are concentrated, but advanced geospatial modeling allows analysts to ask more sophisticated questions about causation, interaction, and intervention. These approaches move beyond static representations to capture how emissions, atmospheric processes, and human activity interact dynamically across urban space. While data and computational requirements increase, many of these methods can be implemented incrementally using open or semi-open datasets.

One important class of models focuses on dispersion and atmospheric dynamics. By integrating wind speed, wind direction, temperature, and topography, analysts can estimate how pollutants spread from emission sources over time. Even simplified dispersion models can reveal why neighborhoods downwind of traffic corridors or industrial zones experience higher exposure than areas closer to the source but better ventilated. Temperature inversions and urban heat islands further modify these patterns, trapping pollutants near the surface and intensifying exposure during specific seasons or times of day (WHO, 2022).

A second set of approaches emphasizes network-based exposure modeling. Rather than treating roads as static features, these methods conceptualize transportation networks as moving emission systems. Exposure can be estimated along commuting routes, bus corridors, or freight pathways, capturing the cumulative effect of repeated travel through polluted environments. This is particularly relevant in cities where long commute times and congestion increase daily exposure far beyond what residential location alone would suggest (Khreis et al., 2020).

Scenario modeling offers a third powerful tool. By modifying input layers—such as reducing traffic volumes, rerouting freight, or introducing low-emission zones—analysts can simulate how policy interventions might alter exposure patterns. These models do not predict the future with certainty, but they allow planners to compare relative outcomes across scenarios, making trade-offs explicit. For example, restricting heavy trucks from a central business district may reduce exposure in one area while increasing it along alternative routes, a spatial consequence that GIS is uniquely suited to reveal (WHO, 2022).

Advanced models increasingly incorporate uncertainty and sensitivity analysis. Satellite-derived pollution estimates, proxy traffic measures, and population datasets all contain error. Explicitly representing uncertainty through ranges, confidence surfaces, or ensemble approaches helps prevent false precision and improves transparency in decision-making. In data-scarce contexts, acknowledging uncertainty is often preferable to withholding analysis altogether (WHO, 2022).

Finally, some of the most ambitious approaches integrate health and socio-economic data to examine cumulative risk. While individual-level health records are rarely available, neighborhood-scale indicators of age, income, housing quality, and baseline health can be combined with pollution exposure surfaces to identify populations facing compounded vulnerability. These models shift the focus from pollution alone to risk as a function of exposure, susceptibility, and capacity to respond—an essential perspective for equitable urban planning (Khreis et al., 2020).

10.4 Getting to Action on Air Pollution and Traffic

Urban air pollution poses a difficult challenge for action because it is deeply embedded in how cities function. Vehicles, industries, and energy systems cannot simply be moved or removed without disrupting lives and businesses. At the same time, the health costs of inaction are substantial and unequally borne across urban populations. Moving from analysis to action therefore requires strategies that acknowledge these trade-offs while using spatial evidence to reduce exposure where it matters most (WHO, 2022).

GIS plays a critical role in this transition. By making exposure visible and place-specific, geospatial analysis helps shift air pollution from an abstract environmental concern to a concrete planning problem. Maps reveal where emissions intersect with vulnerable populations, where interventions will produce the greatest health benefits, and where unintended consequences may arise. Action, in this sense, is not about eliminating pollution entirely, but about reconfiguring urban systems to reduce harm.

10.4.1 Principles for Enhancing Urban Resilience

Several principles emerge consistently from cities that have made progress on air quality. First, target exposure, not just emissions. Citywide emission reductions are important, but health benefits accrue most rapidly when interventions focus on locations where people are most exposed, especially near roads, schools, markets, and dense housing. GIS-based hotspot mapping helps identify these priority zones and avoid dispersing resources too thinly (Khreis et al., 2020).

Second, integrate transport, land use, and public health planning. Air pollution is rarely solvable within a single sector. Transport planners may focus on congestion, while health agencies track disease burden and zoning authorities regulate land use. Geospatial tools allow these domains to be analyzed together, revealing how traffic flows, industrial siting, and residential patterns jointly shape exposure. Cities that institutionalize this integration are better positioned to implement durable solutions (WHO, 2022).

Third, prioritize equity explicitly. Because air pollution exposure follows socio-economic gradients, neutral or market-driven interventions can inadvertently worsen inequality. For example, traffic rerouting may protect affluent neighborhoods while increasing exposure elsewhere. GIS-based equity analysis—overlaying pollution with income, age, or housing quality—helps ensure that interventions reduce rather than redistribute harm (Khreis et al., 2020).

Finally, plan for compound risks. Climate change, wildfire smoke, and heat amplify urban air pollution in ways that exceed historical norms. Resilience requires anticipating these interactions and embedding flexibility into air quality management strategies, rather than relying solely on fixed regulatory thresholds (WHO, 2022).

10.4.2 Plans and Policy Levers

At the policy level, cities have a growing array of options including vehicle emission standards, fuel quality regulations, low-emission zones, congestion pricing, industrial permitting, and land-use controls. The effectiveness of these measures depends not only on legal authority but also on actual implementation. GIS enables planners to test where and how policies should be applied to maximize exposure reduction.

Technological advances have expanded the practical toolkit for urban air quality management. Satellite-based monitoring now provides near-global coverage of key pollutants such as nitrogen dioxide and ozone, allowing cities without dense ground networks to identify broad exposure patterns (WHO, 2022). Low-cost sensors, while less precise, can supplement these data by capturing street-level variation and engaging communities in monitoring efforts.

Smart-city platforms increasingly integrate air quality data with traffic information, weather forecasts, and mobility systems. When designed thoughtfully, these dashboards support real-time decision-making—such as adjusting traffic flows during high-pollution days or issuing targeted public health advisories. GIS acts as the backbone of these systems, linking disparate data streams into a coherent spatial framework.

However, technology alone is not a solution. Poorly calibrated sensors, opaque algorithms, or inaccessible visualizations can undermine trust and obscure uncertainty. Best practice emphasizes transparency, clear communication of limitations, and alignment with planning objectives rather than novelty. In many low- and middle-income cities, simpler approaches, including well-designed maps, seasonal summaries, and scenario comparisons, may be more actionable than real-time systems that exceed institutional capacity (WHO, 2022).

Air pollution also reflects physical design choices. Road width, intersection density, building height, and ventilation corridors all influence pollutant accumulation. Cities can reduce exposure through targeted redesign, including traffic calming, narrowing lanes, introducing street trees where appropriate, preserving airflow corridors, and separating heavy freight routes from residential areas. GIS-based modeling helps evaluate these interventions before construction, identifying where design changes will meaningfully improve air quality rather than merely shifting emissions.

For many urban residents, air pollution is not an environmental abstraction but a daily bodily experience. Street vendors working at busy intersections, traffic police stationed along arterial roads, and bus drivers navigating congested corridors all accumulate high exposure over the course of a workday. Children walking to school along traffic-heavy routes and older adults living near industrial zones face similar risks without occupational protections.

These exposure patterns are shaped by necessity rather than choice. People with limited economic resources are less able to modify behavior, relocate, or access protective technologies such as air filtration. GIS-based exposure mapping makes these constraints visible by linking pollution to time, movement, and livelihood. This perspective reframes air pollution as a labor and justice issue, not simply an environmental one (WHO, 2022).

The most effective air pollution interventions are those that reshape how cities move and function over time. One common strategy is traffic redirection—diverting heavy freight away from central districts, schools, and hospitals. GIS-based network analysis allows planners to evaluate alternative routes and assess trade-offs before implementation.

Another strategy focuses on modal shift. Expanding public transit, dedicated bus lanes, cycling infrastructure, and walkable neighborhoods reduces reliance on private vehicles and lowers aggregate emissions. Cities such as Mexico City demonstrate that sustained investment in multimodal transport can reduce pollution while maintaining mobility, particularly when supported by land-use policies that encourage density near transit corridors (WHO, 2022).

Cities have also experimented with vehicle restriction schemes, such as license-plate-based driving bans or congestion charges. Results vary, but GIS-based evaluation helps distinguish short-term behavioral responses from long-term structural change. Without complementary transit options, restrictions may simply shift traffic patterns rather than reduce exposure.

Longer-term strategies involve urban retrofitting. Most cities are not built from scratch and must adapt incrementally. GIS supports this process by identifying priority blocks, corridors, and districts where small changes—relocated depots, redesigned intersections, or improved public facilities—can yield disproportionate health benefits.

Nature-based solutions offer additional co-benefits. Urban forests, green belts, and even modest roadside vegetation can reduce particulate exposure and mitigate heat, though they are not substitutes for emission reduction. GIS helps locate where such interventions are most effective and where they may interfere with ventilation (WHO, 2022).

10.4.3 Looking Ahead

Getting to action on air pollution requires acknowledging its structural roots while using spatial evidence to reduce harm in the present. Traffic and industry will remain central to urban economies for the foreseeable future, but their impacts need not be accepted as inevitable. By integrating GIS into transport planning, zoning of industrial activities, and metrics for assessing public health impacts, cities can make exposure visible, interventions targeted, and outcomes more equitable.

Air pollution is a reminder that urban hazards are rarely isolated. They emerge from the interaction of infrastructure, environment, and inequality. Addressing them effectively demands tools that can navigate this complexity—tools that geospatial analysis is uniquely positioned to provide.

Air pollution demonstrates how environmental hazards translate into uneven health outcomes across urban space. The next chapter builds on this insight by examining how water, sanitation, and disease risks—like air pollution—are structured by urban infrastructure, uneven exposure, and constrained choice.

References

Anenberg, S. C., Miller, J., Henze, D. K., Minjares, R., & Achakulwisut, P. (2019). A global snapshot of the air pollution–related health impacts of transportation sector emissions in 2010 and 2015. *Environmental Research Letters, 14*(12), 124001. https://iopscience.iop.org/article/10.1088/17489326/ab54f5

Bălă, G. P., Râjnoveanu, R. M., Tudorache, E., Motoc, N. S., Oancea, C., & Tudorache, V. (2021). Air pollution exposure—The (in)visible risk factor for respiratory diseases. Environmental Science and Pollution Research, *28*, 19615–19628. https://doi.org/10.1007/s11356-021-13208-x

Bureau of Transportation Statistics. (2023). *Air pollution trends in selected metropolitan statistical areas*. U.S. Department of Transportation. https://www.bts.gov/content/air-pollution-trends-selected-metropolitan-statistical-areas

Fecht, D., Hansell, A. L., Morley, D., Dajnak, D., Vienneau, D., Beevers, S., Toledano, M. B., Kelly, F. J., & Anderson, H. R. (2019). Spatial and temporal associations of road traffic noise and air pollution with cardiovascular mortality in London. *Environment International, 127*, 41–49. https://doi.org/10.1016/j.envint.2019.03.001

Goldberg, D. L., Anenberg, S. C., Kerr, G. H., Mohegh, A., Lu, Z., & Streets, D. G. (2021). Greater inequality observed in urban air pollution exposure between cities than within cities in the United States. *Proceedings of the National Academy of Sciences, 118*(7), e2021669118. https://doi.org/10.1073/pnas.e2021660118

Khreis, H., Kelly, C., Tate, J., Parslow, R., Lucas, K., & Nieuwenhuijsen, M. (2020). Exposure to traffic-related air pollution and risk of development of childhood asthma: A systematic review and meta-analysis. *Environment International, 134*, 105259. https://doi.org/10.1016/j.envint.2019.105259

Malstead, M. (2022). *Air quality and vehicular emissions* [Senior capstone project, University of Richmond]. University of Richmond Scholarship Repository. https://scholarship.richmond.edu/cgi/viewcontent.cgi?article=1053&context=environmentalstudies-seniorseminar

World Health Organization. (2022). *Air pollution and health*. WHO. https://www.who.int/health-topics/air-pollution

11

Mapping Health

Chapter Introduction

Health outcomes in cities are shaped by the environment as much as by medicine. Air, water, and waste systems form the invisible infrastructure of public health—and when they fail, everyday conditions can rapidly become life-threatening. In many cities around the world, rapid urbanization has outpaced investments in sanitation, drainage, and primary healthcare, concentrating environmental exposures in precisely the places where populations are growing fastest. The result is a geography of risk in which disease burden is unevenly distributed across neighborhoods, seasons, and stages of the life course, reflecting not only biological vulnerability but also historical patterns of infrastructure provision and governance.

Health, in this sense, is neither a static condition nor a purely clinical outcome. It emerges from interactions among environmental systems, social organization, and access to care operating across multiple spatial and temporal scales. A child's risk of diarrheal disease may hinge on household water storage practices, neighborhood drainage, regional rainfall patterns, and the reach of public health surveillance networks. These multiple interaction factors make urban health both difficult to manage and uniquely suited to spatial analysis. By integrating environmental data, infrastructure networks, and population characteristics, geospatial tools help reveal how health risks accumulate, shift, and intensify, often long before they register as medical emergencies. This chapter examines how mapping health, sanitation, and environmental risk can move cities from reactive crisis response toward anticipatory, preventative action.

11.1 Setting the Scene: Health as Environmental Infrastructure

In January 2022, Mount Nyiragongo erupted on the edge of Goma, a city of more than 1 million people in eastern Democratic Republic of Congo. Lava flows destroyed neighborhoods, displaced hundreds of thousands of residents,

 DOI: 10.1201/9781032636948-11

and severed already fragile water and sanitation networks. The eruption occurred amid an ongoing civil conflict, a prolonged refugee crisis linked to regional instability, and a public health system stretched to the breaking point. Within weeks, cholera cases surged along the shores of Lake Kivu, not because the disease was novel, but because the environmental and institutional conditions that keep it at bay had collapsed (WHO, 2022; UN OCHA, 2022).

This was a completely preventable disease outbreak. Cholera is well understood, easily diagnosed, and highly responsive to basic interventions: clean water, adequate sanitation, timely surveillance, and rapid treatment. Yet in Goma, the intersection of natural hazards, forced displacement, weak infrastructure, and political instability transformed a manageable public health risk into a humanitarian emergency. Figure 11.1 shows the provincial capital of Goma, Democratic Republic of Congo, with the active volcano of Mount Nyiragongo in the background. This combination of overlapping shocks exemplifies what humanitarian agencies describe as a complex humanitarian crisis, a situation in which conflict, environmental hazards, displacement, and institutional failure interact across multiple spatial and temporal scales, overwhelming local capacity to respond (UN OCHA, 2019).

Health crises of this kind are not simply medical events. They are spatial failures. Pipes, pumps, treatment plants, clinics, roads, and data systems form an invisible geography of care that determines who is exposed, who receives treatment, and who is left behind. In many cities around the globe,

FIGURE 11.1
Goma DRC photo CC-BY-SA 2.0.

rapid urbanization has outpaced the expansion of sanitation systems, waste management, and primary healthcare. Informal settlements often grow fastest in flood-prone lowlands, or near contaminated water sources or open dumpsites, and that is precisely where environmental exposure and disease risk are highest (UN-Habitat, 2020).

The present-day erosion of global disaster response capacity further compounds these risks. For decades, international public health surveillance, combined with early warning systems and humanitarian logistics, was coordinated through U.S.-supported institutions. These played a central role in outbreak detection and response. The weakening or dismantling of these systems has left critical gaps in disease monitoring, food security surveillance, and emergency coordination, with disproportionate consequences for cities already facing environmental stress and political instability (Gostin & Friedman, 2023).

As depicted in Figure 11.2, the USGS One Health approach addresses the interconnection between people, animals, plants, and their shared environment.

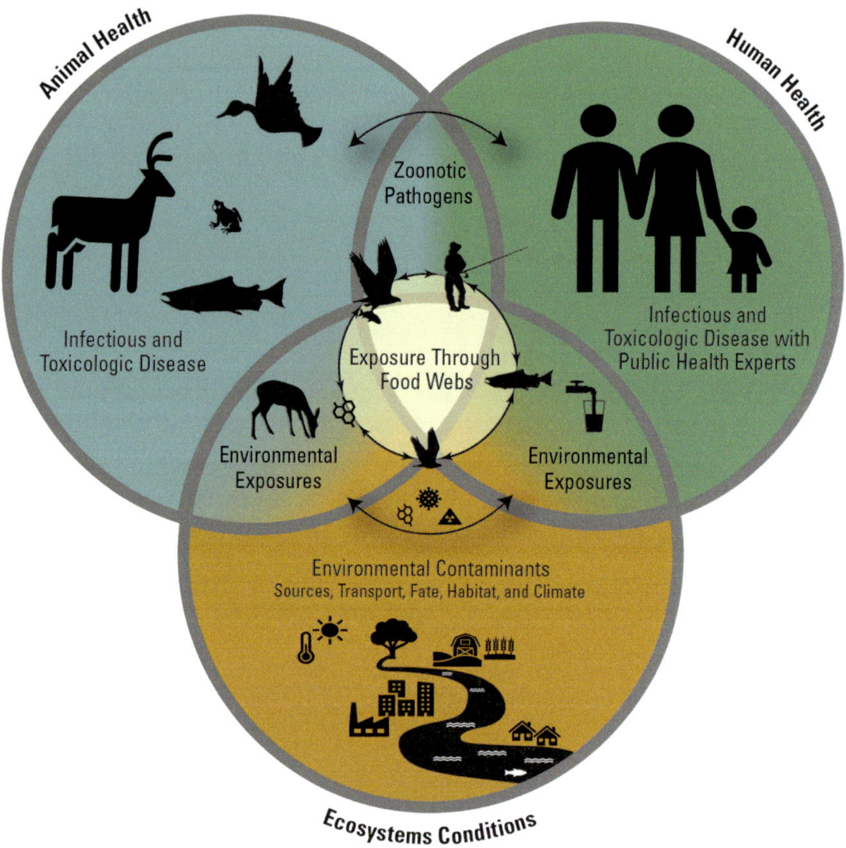

FIGURE 11.2
One health graphic. Public Domain.

This approach helps to answer questions on how and where environmental contaminants originate, how they move through the environment to points of human and wildlife exposure along a source to receptor pathway, and provides an understanding of whether they pose a health hazard for wildlife. The conceptual diagram shows some of the interactions among three compartments: human health, animal health, and environmental conditions.

This chapter adopts a One Health perspective, recognizing that human health is inseparable from environmental conditions and social systems. Health operates across multiple scales—from household water storage practices to regional climate patterns and global supply chains—and across disciplines, linking epidemiology, urban planning, hydrology, governance, and geography. Importantly, health is not synonymous with medical care. Clinics and hospitals matter, but so do clean air, safe water, adequate housing, meaningful work, mobility, and social cohesion.

Geospatial tools offer a way to make these relationships visible. By mapping the spatial distribution of clinics, populations, water infrastructure, vegetation, and disease incidence, GIS helps reveal why outbreaks occur where they do—and how targeted interventions can prevent them. The tragedy of Goma is not only that lives were lost but also that the spatial signatures of risk were legible long before the outbreak began. The task of health mapping is to ensure that such warnings are no longer ignored.

11.2 Unpacking the Threat

Urban health threats across the globe emerge from the tight coupling of environmental systems, infrastructure, and human population. Infectious and non-infectious diseases do not distribute randomly across cities; they cluster where exposure pathways, vulnerability, and institutional weakness overlap. This section focuses on environmentally mediated public health threats, emphasizing how water, sanitation, waste, and settlement patterns structure disease risk in space and time. It can help us learn why these threats remain persistent despite decades of technical knowledge about prevention (WHO, 2023; UN-Habitat, 2020).

11.2.1 The Threat: Environmentally Driven Disease Risk in Cities

Urban public health hazards are most acute where environmental systems intended to protect populations instead become vectors of exposure. In many cities across the globe, waterborne, vector-borne, and waste-related diseases thrive not because pathogens are exotic or poorly understood, but because the basic infrastructures that would interrupt transmission elsewhere are unevenly distributed or intermittently functioning. Cholera, typhoid, dysentery, dengue, malaria, and Zika all exploit predictable environmental conditions, including contaminated water, standing surface pools, poor drainage, unmanaged waste, and dense human settlement (WHO, 2022).

Waterborne diseases remain among the most lethal and preventable urban health threats. Cholera, caused by *Vibrio cholerae*, spreads rapidly where drinking water sources are contaminated by human waste, a condition common in cities lacking a continuous piped supply or effective wastewater treatment. Interruptions in water service often force households to rely on informal sources or storage practices that increase contamination risk, particularly during floods or post-disaster displacement (Ali et al., 2015). These risks are not evenly distributed: informal settlements located in low-lying or flood-prone areas experience repeated exposure as runoff overwhelms drains and latrines, washing pathogens into surface water and shallow wells (UNICEF, 2021).

Vector-borne diseases exhibit a similarly spatial logic. Mosquitoes that transmit dengue, malaria, and Zika breed in standing water created by clogged drains, discarded containers, construction sites, and poorly maintained water infrastructure. Urban expansion often produces fragmented landscapes—paved surfaces adjacent to vegetated patches and open water—which are ideal for vector habitats (Messina et al., 2019). Seasonal rainfall and temperature shifts further modulate transmission, creating predictable spikes in disease incidence that align closely with environmental conditions rather than clinical capacity alone.

Figure 11.3 illustrates the disease ecology of vector-borne viruses through the interaction of mosquitoes, livestock, wildlife, and humans.

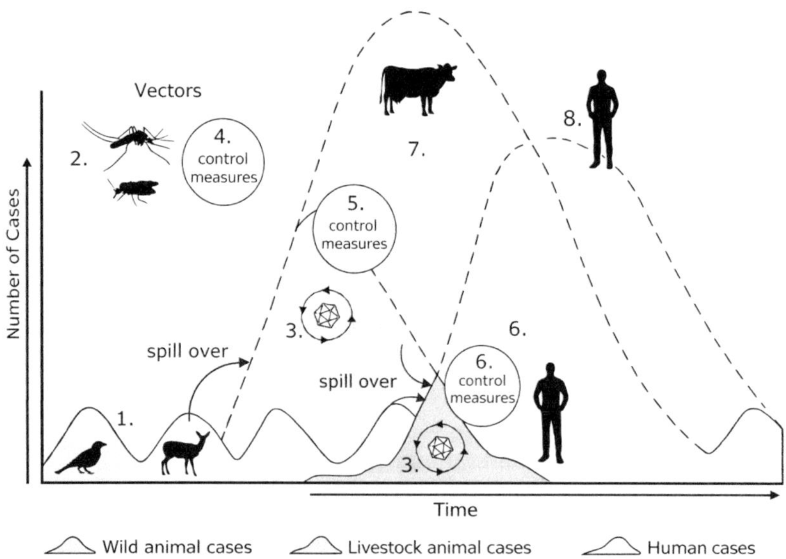

FIGURE 11.3
Disease ecology of vector-borne viruses. (Redrawn from Tim Möhlmann (2019). Mosquitoes, midges and microbiota: European vector diversity and the spread of pathogens.)

Waste-related hazards compound these risks. Open dumpsites generate leachate that contaminates soil and groundwater, while burning waste releases toxic particulates that exacerbate respiratory illness. In dense neighborhoods, waste accumulation blocks drainage channels, increasing flood duration and expanding mosquito breeding grounds. These feedback loops illustrate a key principle of urban health geography: environmental hazards rarely operate in isolation. Instead, they reinforce one another through shared infrastructure failures and land-use patterns (UN-Habitat, 2020).

Importantly, these threats coexist with rising burdens of non-communicable disease. Many cities now face a "double burden" in which infectious diseases persist alongside growing rates of hypertension, diabetes, and respiratory illness linked to air pollution, diet, and stress (GBD 2019 Risk Factors Collaborators, 2020). Health risk therefore accumulates across the life course, with age-specific vulnerabilities layered onto spatially structured environmental exposure. Understanding where and why these risks concentrate is fundamentally a geographic task—one that requires moving beyond aggregate statistics toward spatially explicit analysis.

11.2.2 Visualizing the Threat: Hows and Whys

Many urban health threats become legible only when viewed spatially. Environmental exposure is often localized, repetitive, and visually mundane—yet cumulatively lethal. Maps and images help explain *why* disease risk concentrates where it does by linking everyday urban conditions to biological transmission pathways. Contaminated water sources, standing surface water, unmanaged waste, and polluted indoor air are not isolated problems; they reflect systemic gaps in infrastructure, governance, and service provision. Visualizing these conditions grounds abstract health statistics in the physical landscapes where exposure occurs, revealing how sanitation failures, drainage blockages, and household energy choices translate into predictable patterns of illness. Together, these images illustrate how environmental hazards structure disease risk long before patients enter clinics or outbreaks are formally detected (WHO, 2022; UN-Habitat, 2020).

11.2.3 Global Burden and Damages

Environmentally mediated health threats account for a substantial share of global disease burden, with much of it being preventable. Unsafe water, sanitation, and hygiene are responsible for an estimated 1.4 million deaths annually, primarily from diarrheal diseases, with the heaviest toll borne by children under five (WHO, 2023). Vector-borne diseases add hundreds of millions of cases each year, while household and ambient air pollution together contribute more than six million premature deaths globally (GBD 2019 Risk Factors Collaborators, 2020). These burdens are unevenly distributed across cities, reflecting stark spatial inequalities in infrastructure, exposure, and access to care.

The impacts of environmental health threats in cities extend far beyond individual illness, shaping patterns of vulnerability, mobility, and economic opportunity across urban space. Disease risk is rarely uniform within a city. Instead, it clusters in neighborhoods where environmental exposure overlaps with high population density, insecure tenure, limited services, and weak political representation. Informal settlements, which are often located on floodplains, steep slopes, or marginal land near waste sites, experience a disproportionate share of preventable illness because they are at the intersection of environmental hazard and infrastructural neglect (UN-Habitat, 2020).

Waterborne disease offers a clear example of how urban form and infrastructure shape health outcomes. In cities without continuous piped water, households rely on shared taps, tanker deliveries, or surface water sources. These arrangements increase contamination risk and require time-intensive water collection, disproportionately burdening women and children. During heavy rainfall or flooding, latrines and septic systems overflow and introduce pathogens into streets, homes, and water sources. The result is a spatially patterned cycle of exposure in which the same neighborhoods experience repeated outbreaks following seasonal rains or infrastructure failures (Ali et al., 2015).

Vector-borne diseases exhibit similarly patterned impacts. Dengue, malaria, and Zika thrive in dense urban environments where standing water accumulates in drains, containers, and construction sites. Urban expansion often fragments landscapes in ways that favor mosquito breeding, while inadequate waste management and unreliable water supply encourage household water storage that further increases risk (Messina et al., 2019). These diseases impose recurring health shocks that strain households and clinics alike, particularly during seasonal peaks when caseloads surge simultaneously across multiple neighborhoods.

Environmental health threats also interact with displacement and mobility. Following disasters—floods, earthquakes, volcanic eruptions, or conflict—large populations may be forced into temporary shelters or informal camps with limited water and sanitation. These conditions accelerate disease transmission, as seen repeatedly in cholera outbreaks following mass displacement in cities such as Goma, Port-au-Prince, and Dhaka (WHO, 2022). Even short-term displacement can have lasting health consequences, particularly for children, older adults, and those with pre-existing conditions.

At the same time, many cities face a growing burden of non-communicable disease layered atop persistent infectious threats. Air pollution from traffic, industry, waste burning, and household cooking fuels contributes to chronic respiratory and cardiovascular illness, while urban diets and sedentary lifestyles increase rates of obesity, diabetes, and hypertension. This "double burden" challenges health systems designed to manage acute infections rather than long-term chronic care (GBD 2019 Risk Factors Collaborators, 2020). Importantly, these risks are also spatially structured: neighborhoods exposed to poor air quality, noise, and heat often coincide with those lacking access to green space, safe transport, and preventive care.

The economic impacts of urban health threats further entrench inequality. Illness reduces productivity, increases household expenditures, and can trigger cycles of debt and insecurity. At city scale, outbreaks disrupt labor markets, strain public budgets, and deter investment. Because these costs accumulate in already marginalized areas, environmental health threats reinforce existing spatial inequalities rather than distributing harm evenly across urban populations (World Bank, 2021).

Crucially, many of these impacts are predictable. Seasonal disease peaks align with rainfall patterns; outbreak hotspots correspond to infrastructure gaps; exposure pathways follow drainage networks, waste routes, and settlement patterns. From a geographic perspective, the persistence of urban health crises reflects not a lack of scientific understanding, but a failure to translate spatial knowledge into preventive action. The next sections examine how geospatial tools can help identify these patterns in advance—revealing where intervention is most urgently needed and where modest improvements in infrastructure and access could yield substantial health gains.

AN AFFECTED PLACE: GOMA, DEMOCRATIC REPUBLIC OF THE CONGO

Goma, a rapidly growing city on the northern shore of Lake Kivu, exemplifies how environmental health risks escalate in contexts of overlapping hazard, displacement, and weak infrastructure. Home to more than 1 million residents, Goma sits atop the western branch of the East African Rift, adjacent to the highly active Mount Nyiragongo volcano. The city has experienced repeated volcanic eruptions, earthquakes, armed conflict, and large inflows of displaced populations from eastern DRC and neighboring Rwanda. These overlapping pressures have left water, sanitation, and health systems chronically overstretched (UN OCHA, 2022).

In May 2021 and again in January 2022, eruptions of Mount Nyiragongo destroyed neighborhoods, cut road networks, and displaced hundreds of thousands of people. Lava flows damaged water pipelines and electricity infrastructure, while earthquakes fractured sanitation systems and storage tanks. Displaced residents crowded into temporary shelters, schools, and informal camps, often with limited access to clean water and functioning latrines. In this context, cholera—a disease long endemic to the Lake Kivu region—spread rapidly (WHO, 2022).

Cholera transmission in Goma follows a clear spatial logic. The city's informal settlements are concentrated along the lakeshore and in low-lying areas vulnerable to flooding. Many households rely on untreated lake water, shared taps, or shallow wells, while sanitation infrastructure is sparse or poorly maintained. During periods of heavy rainfall or infrastructure disruption, fecal contamination enters surface water and household storage containers, creating ideal conditions for *V. cholerae* transmission (Ali et al., 2015). Mapping cholera cases alongside water points, drainage networks, and population density reveals persistent hotspots clustered near informal settlements and displacement sites.

What makes Goma particularly instructive is not the novelty of the disease, but the predictability of the outbreak. Cholera risk was well understood, vaccines and treatments were available, and the environmental drivers of transmission were visible in advance. Yet the convergence of political instability, underinvestment in infrastructure, and repeated shocks prevented sustained preventive action. Emergency responses focused on treatment and short-term water provision rather than durable improvements to sanitation and drainage.

Goma illustrates how urban health crises emerge from compound risk rather than single events. Volcanic eruptions did not cause cholera directly; they amplified existing vulnerabilities by disrupting infrastructure, displacing populations, and overwhelming health services. The resulting outbreak was less a failure of medicine than a failure of environmental governance and spatial planning. As climate change increases the frequency of extreme rainfall and displacement across the region, Goma's experience offers a cautionary example of how known risks can escalate into recurring crises when environmental health systems remain fragile.

11.2.4 Increasing Sub-Threat: Flooding, Displacement, and the Amplification of Cholera Risk

Flooding has emerged as a critical amplifier of waterborne disease risk, especially in Global South cities, and particularly when combined with displacement and informal settlement growth. Climate change is increasing the intensity of rainfall events across much of sub-Saharan Africa, South Asia, and Southeast Asia, overwhelming drainage systems that were already undersized or poorly maintained (IPCC, 2022). In dense urban environments, floodwaters mobilize human waste, industrial contaminants, and solid waste, spreading pathogens across neighborhoods and into water supplies.

Cholera transmission is especially sensitive to these dynamics. Flooding compromises sanitation infrastructure by inundating latrines and septic systems, while also disrupting piped water supply, forcing households to rely on unsafe sources or improvised storage. Displaced populations—whether fleeing floods, conflict, or other hazards—often settle in temporary camps or informal areas with minimal water and sanitation services. These conditions dramatically increase exposure to contaminated water and accelerate person-to-person transmission (WHO, 2022).

The spatial overlap of flooding, displacement, and cholera risk is not accidental. Informal settlements are frequently located in flood-prone areas because safer land is unaffordable or politically inaccessible. When floods occur, these same neighborhoods experience the longest water retention, the slowest cleanup, and the least access to emergency services. Mapping flood extent alongside population density and water infrastructure reveals why cholera outbreaks repeatedly recur in the same urban zones following heavy rains (UN-Habitat, 2020).

Displacement further intensifies risk by compressing populations into environments where sanitation thresholds are quickly exceeded. Even short-term crowding can overwhelm water points and latrines, while longer-term displacement erodes social networks and health surveillance systems. In cities such as Goma, Beira, and Dhaka, cholera outbreaks following floods have become cyclical rather than exceptional events (Ali et al., 2015).

From a geospatial perspective, flooding transforms cholera from a localized hazard into a citywide threat by expanding exposure pathways along drainage networks and transportation corridors. These dynamics underscore the need to model health risk as a function of environmental processes and human mobility—not simply disease incidence. As flooding becomes more frequent under climate change, the ability to anticipate where waterborne disease risk will intensify will be central to urban health resilience.

11.2.5 Discussion: From Environmental Exposure to Spatial Risk

The persistence of preventable urban health crises reflects not a lack of medical knowledge, but a failure to anticipate how environmental conditions and population dynamics interact across space and time. Waterborne and vector-borne diseases follow well-established transmission pathways, yet outbreaks recur because these pathways are embedded in urban landscapes shaped by inequality, informality, and uneven service provision. Health outcomes, in this sense, are produced geographically—by where people live, how cities grow, and which neighborhoods receive sustained investment.

Risk from urban health threats accumulates gradually and often invisibly. Contaminated water sources, blocked drains, unmanaged waste, and indoor air pollution are everyday conditions rather than sudden shocks. Their health impacts may remain latent until seasonal rains, flooding, or displacement push already stressed systems beyond critical thresholds. By the time clinics report rising caseloads, environmental exposure has often been widespread for weeks or months. This temporal lag helps explain why responses remain reactive rather than preventative (WHO, 2023).

Geographic scale further complicates intervention. Health risks operate simultaneously at household, neighborhood, city, and regional scales. A household's water storage practices influence exposure, but so do neighborhood drainage networks, citywide sanitation coverage, and regional rainfall patterns. Public health statistics aggregated at municipal or national levels obscure these spatial variations, encouraging broad, undifferentiated responses rather than targeted action where it would yield the greatest benefit (UN-Habitat, 2020).

Geospatial analysis offers a way to bridge this gap by translating environmental and infrastructural conditions into actionable representations of risk. Mapping health facilities, population density, water and sanitation infrastructure, land cover, and mobility networks makes it possible to identify where exposure pathways intersect and where vulnerability concentrates. Importantly, GIS does not replace epidemiology or clinical surveillance; it complements them by situating health outcomes within the physical and social environments that shape transmission.

The examples discussed in this chapter illustrate how known risks escalate into crises when spatial information is fragmented or ignored. In Goma, the alignment of informal settlements, flood-prone terrain, disrupted water systems, and displacement created predictable cholera hotspots. Similar spatial logics underlie dengue outbreaks in South Asian megacities and respiratory illness in neighborhoods exposed to air pollution and indoor biomass burning. In each case, the geography of exposure was legible before health systems were overwhelmed.

Spatial data alone, however, are not sufficient. Ethical concerns around confidentiality, data quality, and political sensitivity complicate the use of fine-scale health information. Person-level disease records are often unavailable or deliberately anonymized, requiring analysts to rely on proxies such as infrastructure coverage, environmental indicators, and modeled accessibility. These constraints demand methodological care, but they do not negate the value of spatial analysis. Instead, they reinforce the importance of transparent, reproducible approaches that prioritize prevention over prediction.

The following section turns from impacts to methods, examining how health threats can be modeled geospatially using open data and accessible analytical techniques. By focusing on infrastructure, environment, and access—rather than individual diagnoses—GIS provides a practical entry point for identifying risk, guiding investment, and reducing the likelihood that preventable diseases continue to define urban life in the twenty-first century.

11.3 Modeling Threat Risk in Geospatial Terms

Modeling health risk in cities requires shifting focus from individual disease cases to the environmental and infrastructural conditions that structure exposure. In low- and middle-income contexts, fine-grained epidemiological data are often incomplete, delayed, or inaccessible due to confidentiality concerns and limited surveillance capacity. Geospatial modeling therefore plays a complementary role, using spatial proxies such as access to water and sanitation, population density, drainage, land cover, and health facility networks, to identify where health risks are most likely to emerge (Cromley & McLafferty, 2012; WHO, 2023).

Rather than predicting who will become ill, spatial models ask where transmission pathways are most likely to intersect with vulnerable populations. This distinction is critical. Environmental health threats are shaped by infrastructure layout, human settlement patterns, and environmental processes that vary systematically across urban space. GIS makes these patterns visible by integrating heterogeneous datasets into a common spatial framework, allowing analysts to compare neighborhoods, prioritize intervention zones, and evaluate trade-offs among competing investments (UN-Habitat, 2020).

The goal of health risk modeling, then, is not diagnostic precision but anticipatory insight, supporting prevention, preparedness, and spatially equitable decision-making.

11.3.1 Conceptual Tools for Spatial Health Modeling

Several conceptual frameworks guide geospatial analysis of urban health risk. Disease ecology emphasizes interactions among host, pathogen, vector, and environment, highlighting how land use, climate, and infrastructure shape transmission (Meade & Earickson, 2005). Exposure pathways trace how contaminants move through water, air, and soil to reach human populations. Accessibility and network theory frame healthcare as a spatial service, shaped by distance, travel time, and connectivity rather than simple facility counts (Guagliardo, 2004). Finally, risk indexing combines hazard, exposure, and vulnerability into composite surfaces that support comparison across neighborhoods while avoiding reliance on person-level health records (Cutter et al., 2003).

11.3.2 Project Design Questions

Designing geospatial health analyses in urban contexts requires navigating several recurring challenges. First, confidentiality and sensitivity limit access to person-level health records, particularly for infectious diseases. Analysts must therefore rely on spatial proxies—such as infrastructure coverage, population density, and environmental conditions—rather than individual diagnoses (WHO, 2023). Second, scale matters: neighborhood-level patterns may be invisible in citywide averages, while overly fine resolution can introduce false precision or ethical risk. Third, model complexity should be proportional to data quality. Simple models that clearly represent exposure pathways often outperform sophisticated techniques built on uncertain inputs. For these reasons, health mapping projects benefit from starting with accessible questions—such as who can reach care, where exposure is concentrated, and how risk varies seasonally—before advancing to more complex modeling (Cromley & McLafferty, 2012).

11.3.3 Lab: Using Open-Source Data to Map Health Access

A practical entry point into spatial health analysis is mapping access to healthcare facilities. Access is a foundational determinant of health outcomes, influencing early diagnosis, treatment, and disease surveillance. In this lab, we use open-source data to model access to hospitals in Presidente Prudente, Sao Paulo State, Brazil, illustrating the difference between simple distance measures and network-based travel analysis.

Data requirements are modest and widely available. The analysis requires (1) a study area boundary, (2) health facility locations, and (3) a road network. Administrative boundaries for Brazil can be obtained from the Humanitarian Data Exchange (HDX). Hospital locations and road networks are available through OpenStreetMap, which provides reasonably complete coverage for many Brazilian cities and can be accessed directly in QGIS using the QuickOSM plugin (OpenStreetMap Contributors, 2024).

The first step is to visualize Euclidean (straight-line) distance to hospitals. Using hospital point locations, buffers or raster distance surfaces can be generated to show areas within fixed distances (e.g., 5, 10, or 20 km). This approach is computationally simple and useful for identifying broad service gaps, particularly in data-scarce environments. However, Euclidean distance assumes uniform travel conditions and ignores barriers such as road layout, terrain, and river crossings.

To better approximate real-world access, the second step uses network-based analysis. In QGIS, the Service Area tool can calculate travel distances along the road network, using hospitals as origin points and roads as the network layer. Service areas can be generated at multiple distance thresholds (e.g., 10, 20, 30, 40, and 50 km), producing polygons that reflect how far residents can travel along existing roads. Compared to straight-line buffers, network service areas often reveal substantial differences in access, particularly in peripheral neighborhoods where road connectivity is limited. The resulting map in Figure 11.4 visualizes the spatial expression of healthcare coverage rather than health outcomes themselves.

The resulting maps visualize the spatial expression of healthcare coverage rather than health outcomes themselves. Attribute tables can be used to distinguish facility types (e.g., general hospitals versus specialized centers), highlighting functional as well as geographic gaps. What these maps do not show is equally important: they do not capture quality of care, wait times, affordability, or health status of the population. Nor do they reflect seasonal disruptions such as flooding that may temporarily sever road access.

Despite these limitations, access mapping provides a powerful foundation for health risk analysis. When combined with population density layers from

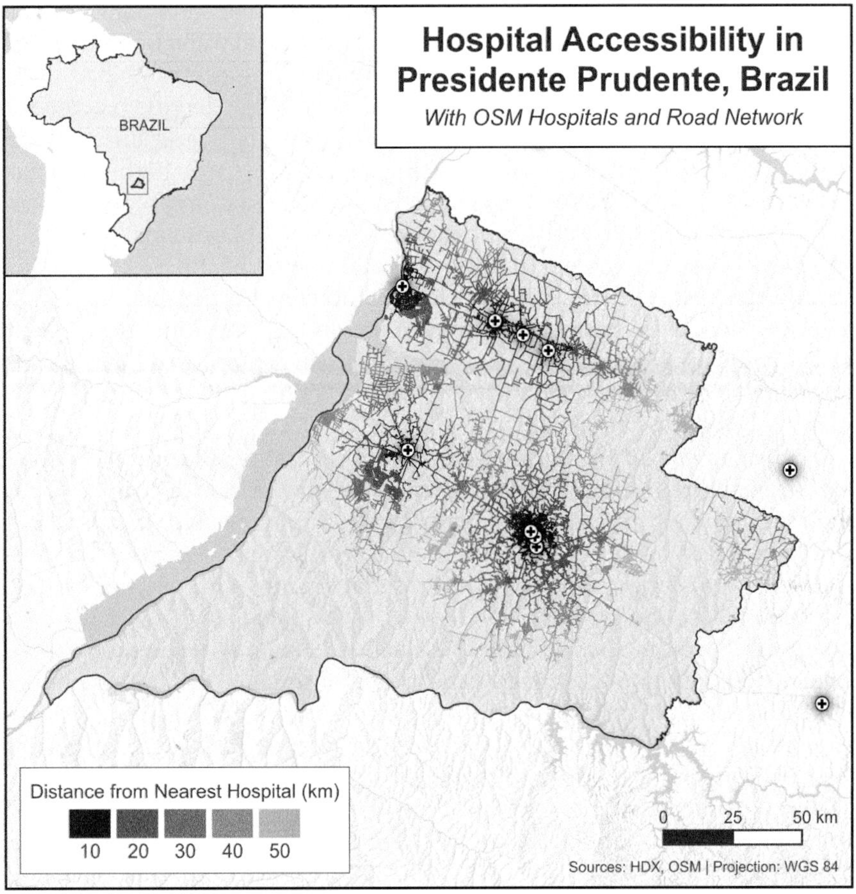

FIGURE 11.4
Brazil healthcare access map. (Cartography by Maya O'Brien.)

WorldPop and environmental indicators such as flooding or air pollution, access surfaces help identify areas where exposure and limited care converge. For cities lacking reliable disease surveillance, this approach offers a transparent, reproducible method for prioritizing investment and emergency preparedness (Guagliardo, 2004; WorldPop, 2023).

11.3.4 Advanced Modeling Approaches

More advanced geospatial health models build on basic access and exposure mapping to represent interaction, uncertainty, and change over time. These approaches are particularly useful in cities facing compound risks, where environmental hazards, infrastructure deficits, and social vulnerability reinforce one another.

One common extension is multivariate risk modeling, in which multiple spatial layers—such as population density, proximity to water bodies, sanitation coverage, vegetation indices, and flood exposure—are combined using regression or weighted overlay techniques. For example, dengue risk models often link mosquito habitat suitability (derived from NDVI and surface water data) with population density and housing quality to identify high-risk zones (Messina et al., 2019). While these models do not predict individual illness, they support comparative risk assessment across neighborhoods.

Scenario analysis allows analysts to explore how health risk may shift under different conditions. Changes in rainfall intensity, population growth, or infrastructure investment can be modeled by adjusting input layers and recalculating risk surfaces. This approach is particularly valuable for climate-sensitive diseases such as cholera, where flooding alters exposure pathways and access to care simultaneously (IPCC, 2022).

Another advanced technique involves incorporating uncertainty explicitly. Data gaps, reporting bias, and temporal mismatch introduce uncertainty into health models, especially in low-resource settings. Sensitivity analysis—testing how results change under different assumptions—helps avoid false confidence and supports more honest communication with decision-makers (Cromley & McLafferty, 2012).

Network-based models can also be extended beyond static access. Travel time models incorporating road quality, slope, or seasonal disruption provide more realistic estimates of service reach. In emergency contexts, dynamic network analysis can simulate evacuation or referral pathways, supporting coordination among clinics and hospitals.

Finally, composite health risk indices integrate hazard, exposure, and vulnerability into a single comparative metric. While such indices inevitably simplify complex realities, they are often effective communication tools, translating multidimensional risk into formats usable by planners and policymakers (Cutter et al., 2003). When grounded in transparent assumptions and updated iteratively, these models help shift health governance from reactive crisis response toward anticipatory planning.

11.4 Getting to Action on Health Threats

Strengthening urban health resilience requires shifting from reactive crisis management toward preventive, spatially informed action. While medical treatment remains essential, most environmental health threats originate outside clinics, in water systems, housing conditions, transport networks, and patterns of land use. Effective intervention therefore depends on principles that recognize health as an outcome of urban systems rather than solely a function of medical care (WHO, 2023).

First, reduce exposure at the source. The most reliable way to lower disease burden is to interrupt exposure pathways before illness occurs. Investments in clean water supply, sanitation, drainage, waste management, and air quality control consistently yield greater population-level health benefits than downstream treatment alone. From a spatial perspective, this means prioritizing neighborhoods where exposure and vulnerability overlap, often informal settlements located in flood-prone or environmentally degraded areas (UN-Habitat, 2020).

Second, improve spatial equity in access to care. Health services must be distributed in ways that reflect population need rather than administrative convenience. Distance, travel time, cost, and reliability all shape whether care is actually accessible. Mapping service areas and referral networks helps identify underserved zones and supports more equitable placement of clinics, mobile health units, and emergency transport capacity (Guagliardo, 2004).

Third, treat health as a multi-scalar system. Health risks operate simultaneously at household, neighborhood, city, and regional scales, shaped by infrastructure, governance, and environmental processes. Household water storage practices interact with neighborhood drainage; citywide sanitation coverage interacts with regional rainfall and watershed dynamics. Policies that focus narrowly on one scale often fail because they ignore these cross-scale interactions. Geospatial analysis helps align interventions across scales, ensuring that local actions are supported by system-level investments (IPCC, 2022).

Fourth, integrate health governance across sectors. Effective health protection depends on coordination among public health agencies, water utilities, housing authorities, transport planners, and environmental regulators. Fragmented governance allows exposure pathways to persist even when individual agencies perform well within their mandates. Spatial data platforms can serve as shared reference points, supporting coordination and accountability across institutional boundaries (World Bank, 2021).

Finally, prioritize transparency and prevention over crisis response. Early warning systems, open data, and routine monitoring of environmental indicators enable cities to act before outbreaks escalate. When health data are delayed, siloed, or politicized, preventable risks become humanitarian emergencies. Resilient cities institutionalize prevention by embedding health considerations into everyday planning decisions rather than treating outbreaks as exceptional events.

Together, these principles reframe urban health as a spatial justice issue—one that can be addressed systematically through informed planning, targeted investment, and sustained attention to the environments in which people live.

11.4.1 Policies and Programs

Translating principles into action requires embedding health objectives into urban planning and infrastructure programs rather than treating them as standalone interventions. City health plans, water and sanitation master plans, climate adaptation strategies, and disaster preparedness frameworks should be spatially aligned and mutually reinforcing. Geospatial analysis helps ensure that investments in clinics, drainage, and sanitation target neighborhoods where environmental exposure and limited access converge (UN-Habitat, 2020; WHO, 2023).

Beyond conventional infrastructure investment, a range of emerging technologies is reshaping how cities monitor, manage, and respond to environmental health threats. Many of these tools complement geospatial analysis by improving data availability, communication, and coordination—particularly in settings where formal surveillance systems are weak.

Mobile health (mHealth) platforms extend the reach of health services by enabling remote consultation, symptom reporting, and appointment scheduling through basic mobile phones. In underserved urban areas, mHealth applications support early detection and continuity of care, especially for maternal health and infectious disease follow-up (Tomlinson et al., 2013). When paired with location data, these systems can also contribute to real-time situational awareness during outbreaks, while still respecting privacy through aggregation and anonymization.

Remote sensing and environmental monitoring provide critical inputs for anticipatory health planning. Satellite-derived rainfall, surface water extent, land surface temperature, and vegetation indices help identify conditions favorable to disease transmission. For example, increases in standing water and vegetation density have been linked to heightened mosquito activity, while flood extent mapping supports cholera risk assessment in low-lying urban areas (Messina et al., 2019; IPCC, 2022). These datasets are increasingly open and updated frequently, making them accessible even to resource-constrained agencies.

Unmanned aerial vehicles (drones) are emerging as tools for both surveillance and service delivery. In some settings, drones have been used to inspect drainage systems, map informal settlements, and deliver medical supplies such as vaccines and diagnostic samples to hard-to-reach neighborhoods. While not a substitute for systemic investment, drones can extend capacity during emergencies or in cities with challenging terrain and limited road access (World Bank, 2021).

Digital health monitoring networks strengthen coordination among clinics, laboratories, and emergency services. Sentinel health networks allow facilities to share anonymized case data, flag unusual disease patterns, and request referrals or transport. When integrated with GIS dashboards, these systems support rapid identification of outbreak hotspots and more efficient allocation of resources (WHO, 2023).

Together, these technologies illustrate that improving urban health does not depend on a single breakthrough. Instead, resilience emerges from integrating modest, well-targeted tools into broader systems of spatial monitoring, infrastructure management, and governance—ensuring that environmental risks are detected early and addressed before they escalate into crises.

11.4.2 Specific Impacts: Infrastructure

Environmental health resilience depends critically on the performance of physical infrastructure systems. Water treatment plants, distribution networks, sanitation facilities, drainage channels, roads, and cold-chain logistics all shape whether health threats are prevented or amplified. Infrastructure failures rarely occur in isolation. Damaged roads delay medical access, broken pipes force unsafe water storage, and inadequate drainage expands vector habitat. From a geospatial perspective, infrastructure investments yield the greatest health gains when targeted to locations where multiple vulnerabilities intersect, such as flood-prone informal settlements with limited clinic access. Mapping infrastructure condition alongside population density and environmental exposure helps prioritize upgrades that reduce disease risk while improving everyday service reliability (UN-Habitat, 2020; World Bank, 2021).

11.4.3 A Human Impact: The Cost of Withdrawing Public Health Capacity

The human consequences of weakened environmental health systems are measured not only in statistics but also in lives disrupted or lost. Public health surveillance, early warning systems, and emergency response networks play a quiet but indispensable role in preventing localized risks from becoming widespread crises. When these systems are dismantled or underfunded, the effects are rarely immediate, but they are profound.

The scaling back of international public health capacity has reduced monitoring of food security, disease outbreaks, and environmental hazards in some of the world's most vulnerable cities. Programs that once tracked rainfall anomalies, crop prices, water quality, and disease incidence provided critical lead time for preventive action. Their absence shifts response from anticipation to improvisation, increasing the likelihood that outbreaks will be detected only after hospitals are overwhelmed (Gostin & Friedman, 2023).

For urban residents living at the margins, these gaps translate directly into harm. Delayed detection of a cholera outbreak means contaminated water sources remain in use longer. Interruptions in vaccine delivery expose children to preventable disease. The erosion of trust and coordination among clinics, utilities, and emergency services further compounds risk, particularly during disasters or displacement. While the full toll of withdrawn public health capacity is difficult to quantify, its effects are visible in the recurrence of preventable outbreaks and the disproportionate burden borne by those with the least ability to adapt.

11.4.4 Looking Ahead: From Healthcare to Healthy Cities

The challenges outlined in this chapter underscore a central lesson: improving urban health requires moving beyond a narrow focus on medical care toward a broader commitment to healthy environments. Clinics and hospitals save lives, but they cannot compensate for contaminated water, unsafe housing, polluted air, or chronic exposure to environmental hazards. Health, understood spatially, is produced through the everyday functioning of cities.

Looking ahead, the most effective health interventions will be those that integrate environmental management, infrastructure investment, and spatial equity. Geospatial tools provide a critical bridge between knowledge and action, enabling cities to identify where risks concentrate, how they change over time, and which interventions are likely to yield the greatest benefit. Importantly, these tools support prevention rather than reaction—helping cities act before exposure escalates into crisis.

There is also a governance dimension to this transition. Health resilience depends on transparency, coordination, and sustained investment. Open data platforms, shared spatial dashboards, and inter-agency collaboration reduce fragmentation and support accountability. When health data are politicized or withheld, environmental risks persist unseen until they erupt into emergencies. Conversely, when cities institutionalize monitoring and early warning, they normalize prevention as part of everyday governance rather than exceptional response (WHO, 2023).

Equity must remain central to this effort. The environmental determinants of health disproportionately affect those living in informal settlements, flood-prone areas, and neighborhoods underserved by infrastructure. Addressing these disparities is not only a moral imperative but also a practical one: reducing exposure in high-risk zones yields outsized health gains at city scale. Mapping makes these inequities visible, transforming abstract commitments to health as a human right into spatially actionable priorities.

Finally, the future of urban health will be shaped by how cities respond to compounding pressures—climate change, migration, and political instability. These forces will test the resilience of health systems and the environments that sustain them. The choice is not whether cities will face new health threats, but whether they will continue to respond reactively or invest in the spatial intelligence needed to anticipate and mitigate risk.

The final chapter turns to that question of possibility. Having mapped danger, it asks how cities can map hope—using geospatial tools not only to diagnose risk but also to design healthier, more resilient urban futures.

References

Ali, M., Nelson, A. R., Lopez, A. L., & Sack, D. A. (2015). Updated global burden of cholera in endemic countries. *PLoS Neglected Tropical Diseases, 9*(6), e0003832. https://doi.org/10.1371/journal.pntd.0003832

Cromley, E. K., & McLafferty, S. L. (2012). *GIS and public health* (2nd ed.). Guilford Press.

Cutter, S. L., Boruff, B. J., & Shirley, W. L. (2003). Social vulnerability to environmental hazards. *Social Science Quarterly, 84*(2), 242–261. https://doi.org/10.1111/1540-6237.8402002

GBD 2019 Risk Factors Collaborators. (2020). Global burden of 87 risk factors in 204 countries and territories, 1990–2019. *The Lancet, 396*(10258), 1223–1249. https://doi.org/10.1016/S0140-6736(20)30752-2

Gostin, L. O., & Friedman, E. A. (2023). Global health law: Responding to new and recurring threats. *JAMA, 329*(3), 193–194. https://doi.org/10.1001/jama.2022.21740

Guagliardo, M. F. (2004). Spatial accessibility of primary care: Concepts, methods and challenges. *International Journal of Health Geographics, 3*(1), 3. https://doi.org/10.1186/1476-072X-3-3

Intergovernmental Panel on Climate Change (IPCC). (2022). *Climate change 2022: Impacts, adaptation and vulnerability*. Cambridge University Press. https://www.ipcc.ch/report/ar6/wg2/

Meade, M. S., & Earickson, R. J. (2005). *Medical geography* (2nd ed.). Guilford Press.

Messina, J. P., Brady, O. J., Pigott, D. M., Golding, N., Kraemer, M. U. G., Scott, T. W., … Hay, S. I. (2019). The current and future global distribution and population at risk of dengue. *Nature Microbiology, 4*(9), 1508–1515. https://doi.org/10.1038/s41564-019-0476-8

OpenStreetMap Contributors. (2024). *Planet dump retrieved from* https://planet.osm.org. https://www.openstreetmap.org

Tomlinson, M., Rotheram-Borus, M. J., Swartz, L., & Tsai, A. C. (2013). Scaling up mHealth: Where is the evidence? *PLoS Medicine, 10*(2), e1001382. https://doi.org/10.1371/journal.pmed.1001382

UN Office for the Coordination of Humanitarian Affairs (UN OCHA). (2019). *Global humanitarian overview*. United Nations.

UN Office for the Coordination of Humanitarian Affairs (UN OCHA). (2022). *Democratic Republic of the Congo: Goma volcanic eruption situation reports*. United Nations.

UN-Habitat. (2020). *World cities report 2020: The value of sustainable urbanization*. United Nations Human Settlements Programme.

UNICEF. (2021). *Drinking water, sanitation and hygiene in urban settings*. United Nations Children's Fund.

World Bank. (2021). *Delivering water, sanitation, and hygiene services in fragile and conflict-affected settings*. World Bank Group.

World Health Organization (WHO). (2022). *Cholera fact sheet*. https://www.who.int/news-room/fact-sheets/detail/cholera

World Health Organization (WHO). (2023). *Global health estimates: Leading causes of death and disability*. https://www.who.int/data/gho

WorldPop. (2023). *Global high-resolution population denominators project*. University of Southampton. https://www.worldpop.org

12

Mapping Hope through Action

Chapter Introduction

This chapter brings the book's central argument to its point of application: geospatial analysis matters not because it diagnoses risk, but because it can be used to change outcomes. Across floods, heat, pollution, landslides, and infrastructure failure, earlier chapters have shown how urban environmental threats are produced spatially, through land use, governance, and unequal investment. This final chapter turns to what follows from that insight. It examines how mapping is used in practice to support resilience, civic engagement, and environmental decision-making, particularly in cities facing complex and compounding risks.

Hope, as used here, is not optimism detached from evidence. It is a spatial condition produced through deliberate choices about where people live, where water is allowed to flow, which ecosystems are protected, and how public resources are allocated. Adaptation and resilience are not abstract policy goals; they are spatial processes that unfold across parcels, neighborhoods, watersheds, and transportation corridors. When cities intervene early—by protecting floodplains, investing in green infrastructure, or redistributing access to safe public space—they do not eliminate risk, but they redesign how it is experienced and by whom.

Rather than offering technical solutions, this chapter shows how geospatial tools already introduced in this book converge in practice. It focuses on how cities, particularly in low- and middle income countries, have translated hazard maps into governance reforms, infrastructure investment, and collective action. In doing so, it returns to the book's opening critique of *death by dashboard*. Mapping becomes most powerful not when it multiplies indicators, but when it clarifies choices, supports accountability, and enables sustained action toward more just and resilient urban futures.

12.1 Setting the Scene: From Disaster to Deliberate Design

You may recall from an earlier chapter the catastrophic landslide that occurred on the outskirts of Freetown, Sierra Leone. Saturated hillsides collapsed without warning, sending a wall of mud, trees, and debris into densely

DOI: 10.1201/9781032636948-12

settled neighborhoods below. Entire homes were buried; roads disappeared; waterways were choked with sediment. More than a thousand people were killed or went missing, and tens of thousands were displaced. The disaster was widely described as "natural," yet its causes were profoundly human, documented to be the deforestation of steep slopes, the uphill creep of informal settlements, and the absence of enforceable land-use controls.

In the days that followed, satellite imagery and aerial photographs circulated globally, revealing the full spatial logic of the tragedy. The landslide scar traced a clear path from degraded uplands to flood-prone lowlands, cutting directly through housing that had expanded into hazard zones over decades. What these images made visible was not only the scale of destruction but also the underlying geography of vulnerability. Risk was not evenly distributed across the city; it was concentrated where governance was weakest, land cheapest, and alternatives fewest.

Yet Freetown's story did not end with emergency response and reconstruction. In the years after the disaster, city leaders, community groups, and international partners began reframing the catastrophe as a planning failure that could be addressed spatially. Hillslope protection zones were delineated using terrain models. Tree-planting initiatives employed local youth to stabilize slopes and restore ecological buffers. Floodplains were reimagined not as land awaiting development, but as critical infrastructure, spaces that absorb water, reduce downstream flooding, and provide public amenities when dry.

What distinguishes Freetown's post-disaster response is not the novelty of its technical tools, since satellite imagery, DEMs, and land-cover data are widely available, but the political decision to act on them. In the years following the Freetown 2017 landslide, hazard mapping shifted from a retrospective- to a forward-looking instrument of governance. Zones of landslide and flood risk were no longer confined to technical reports or donor briefings; they entered public debate, budgetary priorities, and land-use discussions. Mapping became a catalyst rather than a postmortem, an argument for intervention rather than an archive of loss. In this sense, hope emerged not from denying risk, but from confronting it with spatial clarity and political will.

This shift has been closely associated with the leadership of Mayor Yvonne Aki-Sawyerr, first elected in 2018 and re-elected in 2023. Speaking at the World Bank's 8th Urbanization and Poverty Reduction Conference in Washington, DC, in March 2024, Ms. Aki-Sawyerr emphasized the structural challenge facing rapidly growing cities like Freetown: a population of roughly 1.5 million growing at approximately 4 percent annually, projected to reach 2 million by 2028, with nearly 70 percent of residents under the age of 35. Growth, she argued, is not inherently the problem; unmanaged growth is. Local governments, despite being closest to residents and everyday risks, often lack the authority, fiscal capacity, or institutional backing to act proactively. Freetown's experience illustrates how spatial data can help close this gap—by translating risk into legible, defensible claims for land-use control, environmental protection, and investment.

Under the banner of #TransformingFreetown, the program launched in 2019 with 19 targets across 11 priority sectors. It has paired hazard mapping with institutional reform. A modest investment of approximately 1 million dollars was used to digitize tax records and build a cadastral database, expanding the city's revenue base and strengthening its capacity to govern land. Initiatives such as the Tree Tracker app linked reforestation, youth employment, and slope stabilization to spatial monitoring, reinforcing the connection between ecological restoration and risk reduction. Through regional and global networks, including Aki-Sawyerr's role as co-chair of C40 Cities, a group of nearly 100 major cities, led by their mayors, collaborating to tackle the climate crisis by reducing greenhouse gas emissions and building resilient, equitable communities (C40 Cities). Freetown's experience has been framed not as an exception but as a transferable lesson, that resilience emerges when spatial knowledge is embedded in everyday governance. In Freetown, mapping did not merely describe danger; it helped legitimize action, redirect resources, and make long-term safety a matter of public accountability rather than technical discretion.

Freetown's experience mirrors a broader shift underway in cities around the world. From river daylighting projects in Seoul to flood-absorbing parks in Curitiba, disaster has increasingly served as a forcing function for rethinking how urban land is used. These cases demonstrate a central claim of this chapter: that resilience is built not only through engineering but also through deliberate choices about where people live, where water flows, and which spaces cities choose to protect rather than exploit. Mapping, when paired with leadership and participation, becomes a tool for imagining and implementing safer futures.

12.2 Unpacking the Threat Reframed as Hope's Geography

In most chapters of this book, "unpacking the threat" has meant tracing the spatial dynamics of harm, where floods accumulate, heat concentrates, pollution persists, and why certain populations bear disproportionate risk. In this chapter, the analytical lens shifts. The central threat is not a single hazard, but the persistent failure to use spatial knowledge to guide urban decision-making. Where cities ignore land suitability, undervalue ecological systems, or exclude communities from planning processes, risk compounds over time. Conversely, where spatial insight is translated into policy, investment, and design, cities can actively produce safer and more livable futures.

Hope, in this sense, has a geography. It is expressed through protected floodplains, restored wetlands, shaded streets, accessible parks, and infrastructure designed to absorb rather than resist environmental stress. These outcomes are not accidental. They reflect choices about land use, governance authority, and public priorities. Understanding hope's geography therefore requires unpacking how spatial interventions reshape exposure, vulnerability, and opportunity across urban landscapes.

12.2.1 When the "Threat" is Inaction

Urban disasters are often framed as sudden shocks. Storms, fires, heat waves, and landslides arrive often with little advance warning, but their impacts are usually the cumulative result of decades of spatial decisions. Housing placed on steep slopes, highways routed through floodplains, wetlands filled for short-term development gains, and informal settlements pushed into marginal land all represent choices that lock risk into the urban fabric. The true threat, then, lies not only in environmental forces but also in governance systems that fail to anticipate them.

In many cities around the globe, and especially in low- and middle-income areas, rapid urbanization has outpaced formal planning capacity. Yet the absence of perfect data has rarely been the binding constraint. Satellite imagery, elevation models, land-cover classifications, and climate projections are increasingly accessible. What remains uneven is the ability, or willingness, to translate this information into enforceable land-use decisions. Hope emerges where cities close this gap between knowledge and action.

12.2.2 Impacts of Spatial Foresight in Cities

Cities that integrate geospatial analysis into land-use planning consistently demonstrate reductions in disaster losses and co-benefits for public health, mobility, and social cohesion. Flood mitigation projects that restore natural storage capacity—rather than relying solely on concrete channels—reduce downstream peak flows while creating recreational space. Urban greening strategies lower surface temperatures, improve air quality, and enhance mental well-being. These outcomes are spatially uneven, however, unless equity is explicitly built into planning models.

Consider flood management. Traditional approaches often prioritize engineered "gray infrastructure" defenses designed to keep water out. While effective in some contexts, these systems can fail catastrophically when exceeded and frequently transfer risk downstream. By contrast, cities that map floodplains as functional green infrastructure, or spaces where water is allowed to spread safely, reshape the hazard itself. Parks, wetlands, and retention basins absorb floodwaters during extreme events and serve everyday social and ecological functions the rest of the year.

Heat mitigation can follow a similar spatial logic. Urban heat islands emerge from specific land-cover patterns, with dense impervious surfaces, limited tree canopy, and restricted airflow. Mapping these conditions allows cities to target interventions, including tree planting, reflective materials, and cooling centers, where they will have the greatest impact. Importantly, these interventions often overlap with neighborhoods facing social and economic disadvantage, making spatial targeting a matter of environmental justice as much as climate adaptation.

Transportation and land use further illustrate hope's geography. Compact, transit-oriented development reduces emissions, improves access to services, and lowers household transportation costs. When mapped alongside income, age, and disability data, accessibility analyses reveal where investments can most effectively expand opportunity. In these cases, resilience is not a single project but an emergent property of coordinated spatial decisions.

12.2.3 One Affected Place: Daylighting Seoul's Cheonggyecheon Stream

For much of the twentieth century, the Cheonggyecheon stream in central Seoul was treated as an obstacle to modernization. Covered with concrete and eventually buried beneath an elevated highway, the waterway became emblematic of postwar industrial growth. By the late 1990s, however, the structure was aging, congestion was severe, and flooding risks persisted despite extensive engineering. The city's decision to remove the highway and restore the stream marked a radical rethinking of urban space. Guided by hydrological modeling, land-use analysis, and public engagement, planners reintroduced surface water flow, expanded flood storage capacity, and created a linear green corridor through the city center. The project reduced local temperatures, improved air quality, and restored aquatic ecosystems while maintaining flood protection for surrounding neighborhoods.

Crucially, the Cheonggyecheon project reframed water as an asset rather than a liability. Flood risk was managed through space rather than exclusion, and infrastructure investment was leveraged to deliver multiple co-benefits. While Seoul's political and financial capacity is not universally replicable, the underlying spatial logic—daylighting waterways, reclaiming floodplains, and integrating ecological function into urban design—has influenced projects worldwide.

12.2.4 An Increasing Sub-Threat: Competing Claims on Urban Land

As cities pursue resilience-oriented land use, a growing sub-threat emerges, that of the competition for scarce urban land. Floodplains, steep slopes, and coastal margins are often attractive for development precisely because they are under-regulated or historically marginalized. Protecting these spaces can provoke political resistance and displacement concerns, along with unavoidable legal challenges, particularly where tenure is informal.

Without careful governance, adaptation investments risk reinforcing inequality. Green infrastructure can raise property values and trigger displacement unless paired with housing protections. Relocating residents from high-risk zones requires compensation, participation, and viable alternatives. Mapping plays a critical role here—not only in identifying hazard zones but also in documenting social vulnerability, tenure patterns, and access to services.

Cities that succeed in navigating these tensions treat land-use planning as a negotiated process rather than a purely technical one. They use maps to make trade-offs explicit, to justify long-term public benefits, and to build coalitions around shared outcomes. Hope's geography, in this sense, is inseparable from political geography.

12.2.5 Infrastructure as Return on Investment

One of the most compelling arguments for resilience-oriented land use is economic. Studies consistently show that investments in hazard mitigation yield high returns, often exceeding four or five times the initial cost. Floodplain restoration, green roofs, and urban forests reduce damages, lower health expenditures, and extend infrastructure lifespans.

Geospatial analysis strengthens these arguments by quantifying avoided losses and co-benefits spatially. Overlaying hazard maps with asset inventories allows planners to estimate damages prevented under different scenarios. Accessibility models reveal how green space investments improve service reach. These analyses move resilience from aspiration to fiscal strategy.

Importantly, return on investment is not limited to monetary metrics. Reduced mortality, improved mental health, and strengthened social networks represent durable gains that are difficult to price but essential to urban well-being. Mapping helps make these benefits visible and defensible in policy debates.

12.2.6 Green Infrastructure as Flood Protection: Evidence from Recent Research

A growing body of empirical research demonstrates that green infrastructure can deliver measurable, and often superior, flood protection benefits compared to conventional gray infrastructure alone. Recent large-scale analysis published in Scientific Reports provides some of the clearest quantitative evidence to date that nature-based solutions reduce flood impacts while generating substantial co-benefits across urban systems (Scientific Reports, 2025).

Using spatially explicit hydrological modeling across multiple urban catchments, the study shows that interventions such as restored floodplains, urban wetlands, permeable surfaces, and expanded tree canopy significantly reduce peak runoff volumes and delay flood wave arrival during extreme precipitation events. These effects are particularly pronounced in densely developed watersheds, where impervious surfaces otherwise accelerate surface flow and overwhelm drainage infrastructure. Rather than attempting to exclude water entirely, green infrastructure reshapes how water moves through urban landscapes—slowing, storing, and dispersing flows in ways that reduce downstream damage.

Crucially, the study demonstrates that flood mitigation benefits scale nonlinearly with spatial configuration. Distributed green infrastructure networks—small wetlands, linear green corridors, floodable parks, and vegetated streets—often outperform isolated flagship projects. This finding reinforces a central argument of this book: flood risk is not only a function of engineering capacity but also of spatial arrangement. Where green infrastructure is strategically located within upstream catchments, along drainage pathways, and in low-lying zones identified through GIS, cities can achieve disproportionate reductions in flood risk relative to investment cost.

From an economic perspective, the analysis strengthens the case for green infrastructure as a high-return investment. Avoided flood damages, reduced strain on drainage systems, lower maintenance costs, and extended infrastructure lifespans consistently outweighed upfront implementation costs across modeled scenarios. Importantly, these benefits accrued alongside non-flood outcomes: improved water quality, reduced urban heat, enhanced biodiversity, and expanded recreational space. When evaluated spatially, these co-benefits tended to align with neighborhoods historically underserved by infrastructure.

Figure 12.1 illustrates the interactions between forms of green infrastructure, groundwater, and natural ecosystems.

Figure 12.2 displays a bioswale or "rain garden" on a city street. Design improvements such as bioswales can minimize flood damage farther downstream by allowing water to be retained on site longer.

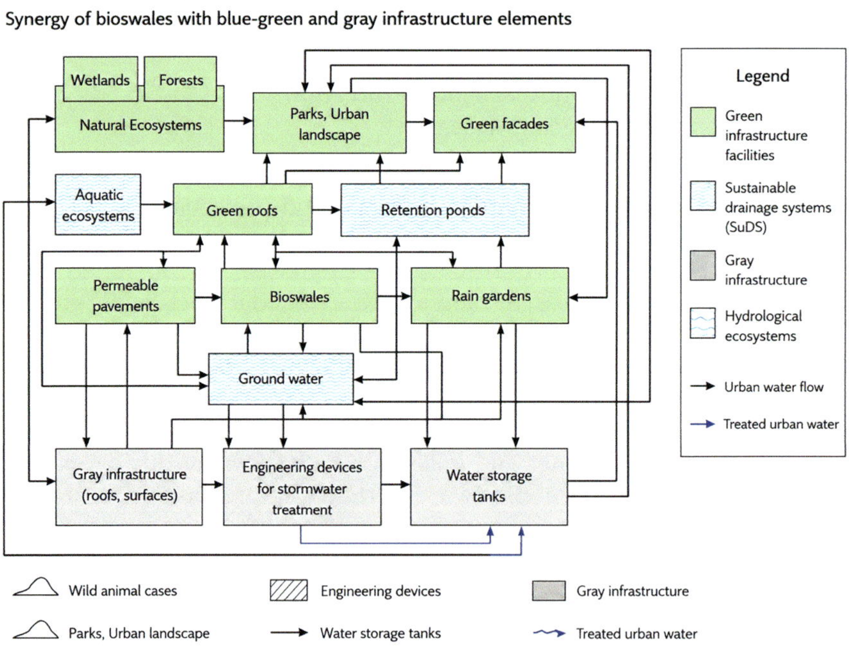

FIGURE 12.1
Blue-green infrastructure. (Redrawn from https://newpolis.media/rain-as-a-resource-how-landscapearchitecture-makes-sustainable-stormwater-management-possible/)

FIGURE 12.2
Urban bioswale in a rainstorm. (Author photo.)

12.2.7 Discussion: From Threat Mapping to Hope Mapping

Reframing "unpacking the threat" as an exploration of hope's geography underscores a central argument of this book: maps do not merely describe urban conditions; they influence how decisions are made. When hazards are mapped without clear pathways to action, spatial knowledge can overwhelm or paralyze. When mapped alongside options for intervention—alternative land uses, investment priorities, and governance choices—those same representations become instruments for change.

Hope, in this sense, does not require the elimination of risk. Risk is an inescapable feature of urban life, particularly in cities shaped by climate variability, rapid growth, and constrained resources. What distinguishes more resilient cities is not their ability to avoid hazard entirely, but their capacity to engage risk deliberately, equitably, and transparently. Mapping plays a critical role in this process by clarifying trade-offs and making the consequences of inaction visible.

Across the examples discussed in this chapter, hope emerges where spatial knowledge is embedded in everyday governance. Floodplains are protected rather than quietly rezoned. Steep slopes are stabilized and reforested instead of incrementally settled. Infrastructure investments are evaluated not only for efficiency but also for whom they protect and whom they expose. In each case, mapping shifts the conversation from whether risk exists to how it should be managed.

Importantly, hope mapping does not substitute optimism for analysis. It relies on the same datasets, methods, and critical scrutiny used to diagnose danger earlier in this book. What changes is orientation. Elevation models, land-cover data, accessibility surfaces, and vulnerability indices are applied prospectively to test alternatives and guide design, rather than retrospectively to explain loss. In this way, mapping becomes less a record of failure and more a tool for foresight.

The geography of hope is therefore neither a fixed outcome nor a single project or policy. It is an ongoing practice that aligns spatial evidence with political commitment and public participation. Cities that sustain this alignment are able to adapt as conditions change, updating maps, revisiting assumptions, and adjusting interventions over time. The task is not to produce perfect representations, but to keep spatial knowledge active within the arenas where choices are made.

This reframing sets the stage for the sections that follow. This chapter now turns from interpretation to design, examining how familiar geospatial techniques can be used to model possibilities, compare interventions, and support decisions that reduce risk while expanding opportunity.

12.2.8 From Mapping Threats to Mapping Possibilities

Throughout this book, geospatial analysis has been used to expose patterns of harm: where floods concentrate, where heat intensifies, where pollution persists, and where infrastructure failure cascades into human loss. These chapters demonstrated that risk is rarely random; it is produced through spatial decisions about land use, infrastructure placement, and political neglect. This final chapter does not abandon that analytical foundation. Instead, it builds directly upon it by asking a different, complementary question: once risks are made visible, what comes next?

Mapping hope requires the same technical rigor as mapping danger, but it applies those tools toward foresight rather than diagnosis alone. The datasets, indicators, and models introduced earlier, including elevation, land cover, accessibility, and vulnerability indices, remain essential. What changes is their purpose. Rather than identifying where cities are most fragile, they are used here to identify where intervention can be most effective, equitable, and durable. In this sense, hope mapping is not a new methodology but a new orientation: one that treats spatial knowledge as a guide for action rather than an endpoint.

12.3 Modeling Hope in Geospatial Terms

Geospatial modeling is often associated with prediction: estimating flood depths, projecting heat exposure, or simulating hazard frequency under climate change. In the context of adaptation and resilience, however, modeling serves a broader function. It allows cities to test alternatives before committing resources, to compare interventions across space, and to anticipate unintended consequences. Modeling hope means asking not only where risk exists but also how it can be reduced through deliberate spatial change.

This section reframes familiar geospatial techniques, including overlay analysis, suitability modeling, accessibility mapping, and scenario analysis, as tools for designing safer and more livable cities. Rather than producing a single "optimal" solution, these approaches support transparent decision-making by revealing trade-offs among competing goals: flood protection, housing supply, ecological integrity, and social equity. The emphasis is on models that are interpretable, participatory, and adaptable across contexts, particularly in cities where data availability and institutional capacity vary widely.

12.3.1 Conceptual Tools for Mapping Hope

Several conceptual tools recur across successful resilience-oriented mapping efforts:

- Suitability analysis to identify land that can safely absorb water, heat, or ecological functions.
- Accessibility modeling to evaluate who benefits from resilience investments and who is left out.
- Ecosystem services mapping to quantify co-benefits such as cooling, air filtration, and flood attenuation.
- Scenario comparison to test alternative futures under different policy or investment choices.

Together, these tools shift mapping from static description to active design, allowing cities to visualize how today's decisions shape tomorrow's risk landscape.

12.3.2 Project Design Questions

When modeling adaptation and resilience, analysts should begin with clear design questions rather than datasets. Typical questions include:

- Where can floodwaters be safely stored without displacing residents or critical infrastructure?
- Which neighborhoods lack access to cooling, green space, or emergency services?
- How would alternative land-use decisions change exposure over time?
- Who benefits—and who bears costs—from proposed interventions?

Framing projects around these questions ensures that models remain policy-relevant and ethically grounded. It also helps prevent "solutionism," where technical outputs outpace institutional or social capacity to act.

12.3.3 Lab: Designing Multi-functional Green Infrastructure with Open Data

This lab introduces a practical workflow for identifying urban spaces that can function simultaneously as green space and floodwater impoundment areas. The goal is not to produce a final design, but to demonstrate how geospatial analysis can guide resilient land-use decisions.

12.3.3.1 Step 1: Assemble Base Layers

Core datasets include a digital elevation model (DEM), land-use or zoning maps, hydrography, and existing green space inventories. Where available, flood hazard maps or historical inundation data can be incorporated. Population density and socioeconomic indicators are essential for evaluating equity implications.

12.3.3.2 Step 2: Identify Physical Suitability

Using slope and elevation thresholds, analysts can exclude areas unsuitable for flood storage or public access. Low-lying zones adjacent to waterways, former industrial land, or underutilized open space often emerge as candidates. Hydrological flow accumulation models help identify where water naturally concentrates.

12.3.3.3 Step 3: Overlay Social Criteria

Accessibility analysis reveals which neighborhoods are underserved by parks or cooling infrastructure. Overlaying vulnerability indicators—such as income, age, or housing quality—helps prioritize areas where co-benefits are greatest.

12.3.3.4 Step 4: Scenario Mapping

Alternative scenarios can be modeled by adjusting assumptions: protecting floodplains versus developing them, expanding green corridors versus isolated parks. Comparing outcomes spatially allows decision-makers to weigh trade-offs explicitly.

This workflow is adaptable across platforms (QGIS or ArcGIS) and scalable from small municipalities to megacities, making it particularly relevant in Global South contexts.

12.3.4 Advanced Modeling Approaches

More advanced applications extend beyond suitability analysis to explore system dynamics and uncertainty. Hydrodynamic models simulate how restored floodplains alter downstream flows. Urban climate models estimate cooling effects of tree canopy expansion under different climate scenarios. Network analyses evaluate how resilience investments affect access to hospitals, shelters, or transit during extreme events.

Scenario analysis is especially powerful for policy discussions. By visualizing outcomes under different investment strategies, cities can move debates from ideology to evidence. Importantly, advanced models should be used judiciously; complexity must serve clarity, not obscure it.

12.3.5 Economic and Institutional Dimensions

Modeling hope also requires attention to governance. Land-use authority, property rights, and funding mechanisms vary widely across cities, shaping what interventions are feasible. Incorporating institutional constraints into spatial analysis, such as jurisdictional boundaries or land tenure patterns, helps align models with real-world decision-making.

Economic analysis further strengthens resilience planning. Cost–benefit assessments that incorporate avoided damages and co-benefits can justify investments in green infrastructure and land protection. When paired with maps, these analyses become powerful tools for public communication and coalition building.

Modeling hope is not about predicting a single future; it is about expanding the range of possible futures that cities can imagine and choose from. By reframing geospatial tools as instruments of design rather than diagnosis alone, this section prepares the ground for the chapter's final turn: translating spatial insight into sustained action.

12.4 Getting to Action on Hope

Mapping hope only matters if it changes decisions. Throughout this book, geospatial tools have revealed where risks concentrate and why. This final section addresses the harder question: how those insights move from maps into budgets, regulations, and lived experience. Action requires more than technical clarity. It depends on authority, incentives, public participation, and persistence, often in contexts where institutional capacity is uneven and political power contested.

This section focuses on how cities convert spatial analysis into durable action. It emphasizes land-use governance, infrastructure investment, and civic engagement as mutually reinforcing components of resilience. While examples vary across regions, the underlying lesson is consistent: resilience emerges where spatial evidence is embedded in everyday decision-making.

12.4.1 Principles for Enhancing Resilience

Across cities that have successfully reduced risk through adaptation, several shared principles recur.

First, prioritize land-use decisions over structural defenses alone.

Infrastructure can reduce risk, but land-use choices determine exposure. Prohibiting settlement in floodplains, steep slopes, and coastal surge zones often delivers greater long-term protection than any engineered solution. Where relocation is necessary, it must be planned, compensated, and participatory.

Second, design for multiple benefits.

Resilience investments that serve only a single function are politically fragile. Projects that combine flood control with recreation, heat mitigation with public space, or water storage with ecosystem restoration are more likely to attract funding and public support. Mapping co-benefits spatially helps justify these investments.

Third, center equity explicitly.

Without intentional design, adaptation can reinforce inequality. Spatial models should therefore integrate social vulnerability, tenure status, and service access—not as afterthoughts, but as core criteria. Equitable resilience is not a moral add-on; it is a prerequisite for political durability.

Fourth, plan for uncertainty.

Climate projections, demographic trends, and economic conditions are uncertain. Resilient systems are flexible rather than optimized for a single scenario. Scenario mapping helps cities compare strategies across plausible futures instead of committing to brittle solutions.

12.4.2 Authority and the Question of Power

Cities vary widely in their authority to regulate land use. In some contexts, municipal governments control zoning and building codes; in others, authority rests with national ministries or provincial governments. Effective action requires aligning spatial analysis with the actual locus of decision-making. Mapping alone cannot substitute for governance, but it can strengthen the case for reform.

12.4.3 Technology Beyond GIS: From Analysis to Accountability

While GIS anchors resilience planning, other technologies increasingly shape how maps influence action. Open-data portals allow residents, journalists, and advocacy groups to scrutinize land-use decisions. Mobile mapping platforms such as OpenStreetMap and Mapillary enable communities to document infrastructure conditions and environmental hazards directly. Citizen-sensing initiatives using low-cost air quality or temperature sensors supplement official monitoring networks, particularly in underserved areas.

These tools shift mapping from a closed technical exercise to a public process. When residents can see how proposed developments intersect with flood zones or heat hotspots, maps become instruments of accountability. This transparency can be uncomfortable for decision-makers, but it is essential for trust.

Importantly, technology alone does not guarantee inclusion. Digital divides persist, and participatory mapping must be designed to avoid privileging already empowered voices. Successful initiatives pair digital tools with workshops, community meetings, and local intermediaries who translate technical outputs into accessible narratives.

12.4.4 Infrastructure as Intervention: Acting on the Built Environment

Infrastructure remains a central lever for resilience, but its design and placement matter as much as its scale. Roads elevated through floodplains can exacerbate downstream flooding; poorly ventilated housing can amplify heat risk. Spatial analysis helps identify where infrastructure investments reduce risk versus where they merely displace it.

Cities that align infrastructure planning with hazard maps consistently reduce long-term losses. Equally important, maintenance and governance determine performance over time. Infrastructure that cannot be operated, enforced, or maintained ultimately reproduces vulnerability.

12.4.5 Leaving: Managed Retreat and the Politics of Relocation

In some locations, risk cannot be mitigated in place. Coastal erosion, recurrent flooding, and unstable slopes make continued occupation untenable. Managed retreat, defined as the deliberate relocation of people and infrastructure away from high-risk zones, remains one of the most politically challenging adaptation strategies.

Mapping plays a critical role here by clarifying where retreat is necessary, where it is optional, and where investment remains viable. Equally important, maps can document who bears the costs of relocation and who benefits from cleared land. Without transparency, retreat risks becoming displacement by another name.

Successful programs treat relocation as a social process rather than a technical fix. Compensation, tenure security in new locations, livelihood access, and community continuity must be planned spatially. Where retreat is paired with land restoration—such as converting vacated floodplains into wetlands or parks—cities can transform loss into shared benefit.

12.4.6 Memorializing: Remembering Risk to Prevent Its Return

Cities often rebuild quickly after disasters, sometimes erasing physical traces of loss. Yet forgetting risk can invite its return. Memorials, preserved hazard markers, and interpretive landscapes serve an important function in resilience: they keep collective memory alive.

Spatial memorialization—marking flood heights, landslide scars, or former shorelines—anchors abstract risk in place. These interventions do not prevent hazards directly, but they shape public understanding and political will. When risk is visible, the pressure to rebuild unsafely diminishes.

In this sense, memorials are part of the resilience infrastructure. They link past experience to future planning, reinforcing the social learning necessary for long-term adaptation.

12.4.7 Lessons for Action: How These Pieces Fit Together

The transition from mapping danger to mapping hope does not follow a single pathway. Yet several lessons emerge consistently across cities.

First, land-use planning is foundational. Infrastructure, emergency response, and social programs all depend on where development is permitted and protected. Maps that inform zoning, permitting, and investment priorities deliver the greatest long-term impact.

Second, coalitions matter. Technical staff, elected officials, community organizations, and external funders must align around shared spatial narratives. Maps serve as boundary objects—tools that different actors can interpret and use without requiring full consensus.

Third, financing follows clarity. Funders respond to well-defined projects with demonstrated co-benefits and quantified impacts. Spatial analysis strengthens cost–benefit arguments and helps cities compete for limited resources.

Finally, participation sustains progress. Resilience is not a one-time achievement but an ongoing practice. Maps that are revisited, updated, and debated keep adaptation embedded in civic life rather than confined to crisis moments.

Table 12.1 lists some options for obtaining funding for urban resilience projects.

The table illustrates how spatial analysis supports investment decisions across scales, linking hazard mapping and vulnerability assessment to concrete funding mechanisms.

TABLE 12.1

Funding Pathways for Geospatially Informed Urban Resilience

Funding Source	Scale	Typical Project Types	Role of Geospatial Analysis
World Bank (IBRD/IDA)	National/city	Flood protection, urban upgrading, transport resilience, land administration	Hazard mapping, cost–benefit analysis, targeting vulnerable populations
Global facility for disaster reduction and recovery (GFDRR)	City/national	Disaster risk reduction, early warning systems, post-disaster recovery	Risk maps, exposure analysis, scenario modeling
Green climate fund (GCF)	National/regional	Climate adaptation, ecosystem restoration, resilience infrastructure	Climate risk mapping, vulnerability assessments, monitoring and evaluation
Adaptation fund	City/national	Floodplain restoration, urban drainage, community-based adaptation	Spatial targeting, beneficiary mapping, co-benefit analysis
C40 cities climate leadership group	City network	Urban climate action planning, peer learning, technical assistance	Emissions mapping, heat/flood risk analysis, implementation tracking
Rockefeller foundation (e.g., 100RC legacy)	City	Urban resilience strategies, governance reform, data platforms	Resilience diagnostics, systems mapping, dashboard-to-action translation
Bloomberg philanthropies	City	Data-driven governance, climate and health, GIS capacity building	Open data portals, spatial performance metrics
**USAID/GIZ	Local/national	Urban services, climate adaptation, land-use planning	Participatory mapping, spatial decision support
National Governments	National	Disaster mitigation, infrastructure, housing	Statutory hazard maps, zoning enforcement

(Continued)

TABLE 12.1 (*Continued*)

Funding Pathways for Geospatially Informed Urban Resilience

Funding Source	Scale	Typical Project Types	Role of Geospatial Analysis
Municipal capital budgets and bonds	City	Green infrastructure, drainage, parks, transport	Project prioritization, avoided-loss modeling
NGOs and multilaterals (e.g., UNDP, UN-Habitat)	Community/ city	Informal settlement upgrading, community resilience	Participatory GIS, vulnerability mapping

** As of July 1st, 2025, USAID has been dismantled,

12.4.8 From Skills to Stewardship

Getting to action is ultimately about stewardship of land, water, infrastructure, and collective futures. The skills developed throughout this book equip readers not only to analyze risk but also to participate meaningfully in public processes. This can include attending hearings, contributing data, challenging unsafe development, and advocating for equitable investment.

Whether the issue is a proposed highway expansion, a new housing development, or the protection of a floodplain, there is almost always a public process through which spatial evidence can be introduced. The hope of this book is that readers will use these tools not only as analysts but also as engaged citizens, turning maps into arguments and arguments into change.

12.4.9 Conclusion: From Death by Dashboard to Mapping Hope

This book opened with a warning. In an era of abundant data, cities risk drowning in dashboards: collections of indicators that document crisis without enabling response. Maps, charts, and metrics proliferate, yet vulnerability persists. The problem has never been information itself, but detachment—knowledge severed from decision-making, from land-use authority, and from the lived realities of people navigating risk every day. *Death by dashboard* describes a condition in which insight accumulates while agency erodes.

This final chapter has argued for a different relationship between spatial knowledge and action. Mapping hope does not replace realism with optimism, nor does it minimize the severity of urban environmental threats. It restores the link between analysis and choice. Throughout this book, hazards have been rendered visible: floodplains traced, heat islands mapped, pollution plumes modeled, and infrastructure failures anticipated. What ultimately determines outcomes, however, is whether these representations enter the arenas where decisions are made.

Those arenas are concrete and contested. They include zoning boards, city councils, land registries, infrastructure agencies, and public planning meetings. It is there that maps either remain technical artifacts or become arguments—tools that clarify trade-offs, legitimize claims, and shape investment. When spatial evidence is absent from these spaces, risk is reproduced by default. When it is present, debated, and acted upon, cities gain the capacity to redesign how risk is distributed.

Hope, in this sense, is neither abstract nor passive. It is produced through deliberate spatial decisions: protecting floodplains rather than building on them; investing in green infrastructure that absorbs heat and water; relocating people from places that cannot be made safe; and ensuring that the benefits of adaptation accrue first to those most exposed. These actions require technical competence, but they depend equally on participation, governance, and political will. Maps alone cannot save lives, but they can make better futures defensible.

For readers—students, analysts, planners, and advocates alike—this book has aimed to offer more than methods. It has presented a way of thinking spatially about risk as a social process. The tools introduced here are intentionally accessible and adaptable, because resilience is not the exclusive domain of experts. It unfolds through public hearings, community mapping efforts, inter-agency negotiations, and contested land-use decisions. In these settings, maps function as boundary objects: they allow diverse actors to engage with shared evidence without requiring full consensus.

The call to action is therefore straightforward. Use maps not only to diagnose danger but also to imagine alternatives. Bring spatial evidence into the rooms where land-use decisions are made. Question development if proposed in known hazard zones. Contribute data where official records fall short. Build coalitions that translate analysis into durable change. Above all, resist the temptation to treat mapping as an endpoint. The purpose of danger mapping is not perfect representation, but transformation.

If dashboards represent a politics of observation, mapping hope represents a politics of engagement. It insists that cities are not merely sites of accumulated risk, but arenas of choice. In that distinction lies the possibility of safer, more just, and more livable urban futures.

12.4.10 One-Page Visual Synthesis

See Table 12.2.

TABLE 12.2

From Hazard Mapping to Hope Mapping

Urban Hazard/ Challenge	Key Spatial Tool	Primary Intervention	Core Benefits
Flooding (riverine and pluvial)	DEMs; floodplain mapping; hydrologic modeling	Floodplain protection; floodable parks; wetland restoration	Reduced flood damages; water storage; recreation; biodiversity
Extreme heat (UHI)	Land-cover analysis; NDVI; temperature surfaces	Urban greening; tree canopy expansion; cooling centers	Lower mortality; improved health; energy savings

(*Continued*)

TABLE 12.2 (*Continued*)

From Hazard Mapping to Hope Mapping

Urban Hazard/ Challenge	Key Spatial Tool	Primary Intervention	Core Benefits
Air pollution	Emissions mapping; sensor networks; exposure analysis	Traffic calming; zoning reform; green buffers	Improved respiratory health; equity gains
Landslides	Slope analysis; land-cover change detection	No-build zones; reforestation; relocation	Reduced fatalities; ecosystem recovery
Water scarcity	Network analysis; demand mapping	Surface storage; watershed protection; reuse	Supply reliability; ecosystem services
Infrastructure failure	Asset mapping; accessibility modeling	Strategic reinvestment; redundancy planning	Service continuity; economic resilience
Inequitable access to green space	Accessibility and equity mapping	New parks in underserved areas	Mental health; social cohesion
Climate displacement risk	Scenario modeling; tenure mapping	Managed retreat with compensation	Reduced long-term loss; social stability

References

Ahern, J. (2011). From fail-safe to safe-to-fail: Sustainability and resilience in the new urban world. *Landscape and Urban Planning, 100*(4), 341–343. https://doi.org/10.1016/j.landurbplan.2011.02.021

Beatley, T. (2014). *Blue urbanism: Exploring connections between cities and oceans*. Island Press.

Bulkeley, H., Edwards, G. A. S., & Fuller, S. (2014). Contesting climate justice in the city. *Antipode, 46*(4), 921–941. https://doi.org/10.1111/anti.12052

C40 Cities Climate Leadership Group. (2023). *Financing urban climate resilience*. https://www.c40.org

C40 Cities Climate Leadership Group. (2021). *Urban climate adaptation and resilience*. C40.

Global Facility for Disaster Reduction and Recovery (GFDRR). (2021). *Resilient cities: Shaping resilient urban futures*. World Bank.

Kabisch, N., Korn, H., Stadler, J., & Bonn, A. (2017). *Nature-based solutions to climate change adaptation in urban areas*. Springer.

Meerow, S., Newell, J. P., & Stults, M. (2016). Defining urban resilience: A review. *Landscape and Urban Planning, 147*, 38–49. https://doi.org/10.1016/j.landurbplan.2015.11.011

Newman, P., Beatley, T., & Boyer, H. (2017). *Resilient cities: Overcoming fossil fuel dependence*. Island Press.

Opperman, J. J., et al. (2009). *Sustainable floodplains through large-scale reconnection. Science, 326*(5959), 1487–1488.

Scientific Reports. (2025). Green infrastructure site prioritization to improve urban flood resilience in Monterrey and Brussels using a decision support model. *Nature Research*. https://www.nature.com/articles/s41598-025-94851-z
United Nations Office for Disaster Risk Reduction (UNDRR). (2015). *Sendai framework for disaster risk reduction 2015–2030*. United Nations.
World Bank. (2021). *Lifelines: The resilience of infrastructure systems*. World Bank Group.
World Health Organization. (2017). *Urban green spaces and health: A review of evidence*. WHO Regional Office for Europe.

Index

Note: **Bold** page numbers refer to tables and *italic* page numbers refer to figures.

V